D0209774

UNIQUE

UNIQUE

The New Science of
Human Individuality

DAVID J. LINDEN

BASIC BOOKS
New York

Basic Books
Hachette Book Group
1290 Avenue of the Americas, New York, NY 10104
www.basicbooks.com

Printed in the United States of America

First Edition: September 2020

Published by Basic Books, an imprint of Perseus Books, LLC, a subsidiary of Hachette Book Group, Inc. The Basic Books name and logo is a trademark of the Hachette Book Group.

The Hachette Speakers Bureau provides a wide range of authors for speaking events. To find out more, go to www.hachettespeakersbureau.com or call (866) 376-6591.

The publisher is not responsible for websites (or their content) that are not owned by the publisher.

Print book interior design by Jeff Williams.

Library of Congress Cataloging-in-Publication Data

Names: Linden, David J., 1961– author.
Title: Unique: the new science of human individuality / David J. Linden.
Description: First Edition. | New York: Basic Books, 2020. | Includes bibliographical references and index.
Identifiers: LCCN 2020024904 | ISBN 9781541698888 (hardcover) | ISBN 9781541698871 (ebook)
Subjects: LCSH: Individuality. | Neurophysiology. | Neurobiology. | MESH: Heredity, Human.
Classification: LCC BF697 .L53566 2020 | DDC 155.2/2—dc23
LC record available at https://lccn.loc.gov/2020024904

ISBNs: 978-1-5416-9888-8 (hardcover), 978-1-5416-9887-1 (ebook)

LSC-C

10 9 8 7 6 5 4 3 2 1

For Jacob and Natalie

Even with insects—
some can sing,
some can't.

KOBAYASHI ISSA
(translated by Robert Hass)

Contents

UNIQUE

Prologue

I SHOULD JUST LEARN TO RELAX AND ENJOY MUSIC, BUT I can't leave well enough alone. Case in point: I'm blasting down the highway on a fine sunny day when Juvenile comes on the car radio with that infectious New Orleans bounce and I start to bop in my seat.

Oh where she get her eyes from? She get it from her mamma!
Oh where she get her thighs from? She get it from her mamma!
Where she learn to cook from? She get it from her mamma!

I'm singing along now—doing the call-and-response and smacking the steering wheel to the beat. But on a parallel track my mind is already chewing on the lyrics. It's the curse of the geek to overanalyze. I start thinking about DNA. OK, she got her eyes entirely from her mamma's and papa's genes, but those thighs? That's probably a mixture of genes and learned eating habits. The population of bacteria resident in her gut affects her metabolism and hence the thickness of her thighs. Her mamma probably taught her how to cook, so that's down to social experience. And we know from identical-twin studies that individual food preferences have only a small genetic component, so not much mamma (or papa) there. But perhaps she inherited the gene variant that confers supersensitivity to bitter foods. So her cooking style, which reflects her food preferences, is likely to be more complicated than the thigh situation. My train of thought only gets more convoluted as the song continues.

> *Why she swear that she the boss? She get it from her mamma!*
> *Why she always gotta call the law? She get it from her mamma!*

Where does her assertive nature come from? Was it how she was raised? Or perhaps the crucial influence of her peers? Have genes contributed to her confidence? There's evidence for this view as well—variants in neurotransmitters and all that. Will she always feel like the boss and have the gumption to say so, or is this confidence just reflective of her present stage of life? We know that personality traits are somewhat changeable in children but are fairly stable in adults in the absence of major trauma.

Yes, I know that, to a large degree, I'm missing the point. Juvenile is not rapping to detail the experiences, developmental randomness, and genetic factors that shape us as individuals. Nonetheless, he raises many of the central issues of human individuality. The protagonist has a list of traits that we learn about as the song unfolds: in addition to being attractive, confident, and skilled in the kitchen, she's funny and she's close to her friends. How did she get that way?

AS A BIOLOGIST, I go to great lengths to minimize individual differences in my experiments. In my lab, we study strains of mice that have been inbred to be as genetically similar to one another as possible. Then, to further reduce variability, we take scrupulous measures to ensure that they are raised in the same boring lab conditions. We typically measure large populations of animals and then take the average of those measures to test our hypotheses. In order to see the central trend, we ignore the outliers. It's a reasonable approach if one is trying to understand those aspects of biology that people (or mice) have in common. But it's far from the whole story.

If I open up a box of lab mice, fresh from the breeder, I can see that they share certain traits. For example, they will all attempt to hide from bright light, they will all stand stockstill in fear when they smell fox urine, and they will all reject drinking water that's been made bitter with quinine. However, it doesn't take much observation to reveal important individual differences. Some mice are more aggressive toward one another and toward my grasping hand. In the absence of threat, some will race around the cage while others sit there calmly. Individual differences can also be found in physiological measures, such as resting stress hormone levels or sleep patterns or the amount of time it takes for food to pass through the digestive system.

How do they get that way?

FOR SEVERAL YEARS, NOT so long ago, I spent quite a bit of time on the dating site OkCupid, looking for my perfect match. For me, online dating was a fascinating, frustrating, and ultimately serendipitous process: I met my wonderful wife on that website. It turns out that meeting one's future spouse online is not an unusual event these days. According to OkCupid cofounder Christian Rudder, in 2013 there were approximately thirty thousand first dates every night because of his website. Of those pairs, about three thousand became long-term couples and about two hundred eventually married (and presumably many others are in long-term, committed but unmarried relationships).[1] One can only imagine that these numbers have increased significantly since 2013, and of course OkCupid is just one of many such sites.

For me, browsing dating profiles on OkCupid was a master class in human individuality. You probably know how it works. Each person provides basic information, photos, and answers to a set of questions in order to express who they are—or what version of themselves they imagine to be most appealing—and what sort of relationship they're seeking. Importantly, a dating website is not an entirely public forum. Unlike a bar or some other real-world social space, the OkCupid user is not being observed by her friends and coworkers as she writes her profile. She can express herself in a way that's less constrained by social pressures (or at least constrained by a different set of social pressures). Here's a fictitious yet plausible example from my own urban, midlife demographic. Each line is from a real profile, but I've mixed many profiles together to create this composite.

CharmCitySweetie, 54

Woman, straight, single, 5'10", curvy

White, speaks English and Spanish, graduated from university

Somewhat lapsed Roman Catholic, Scorpio

Never smokes, drinks socially, has dogs

Has kids, but does not want more

Seeking single men for long-term dating

My self-summary:

I'm a typical eldest child from a big family.

I love witty banter, wry humor, and self-deprecation.

I work as a defense attorney.

I'm left-handed and proud of it.

I like to walk around the house with my toothbrush in my mouth.

I have a glue gun, and I'm not afraid to use it.

I've been known to get a little blue in the winter.

I like to get up really early and run with my dogs.

I'm always turning the thermostat up.

Both my mind and my browser have too many tabs open at the same time.

I like to sing and I have perfect pitch.

I pick my teeth with the mail when I think nobody is looking.

The first things people usually notice about me:

My long red hair.

The tattoo of a dancing platypus on my shoulder.

My Boston accent.

My perfect memory for song lyrics.

Favorite books, movies, shows, and food:

I don't watch much TV, but I love scary movies. Right now, I'm reading and enjoying the new Jennifer Egan novel. I rock out to everything from sweaty punk to 70s soul music to Schubert. I like spicy food and hoppy beer but despise mayonnaise, mustard, and runny eggs. Red wine is not optional for my happiness.

You should message me if:

You're not standing in front of your boat, toilet, motorcycle, or car in your profile picture.

You know what a "mensch" is and you are one.

You won't mind my cold feet on your back.

You can bring it.

Like the protagonist of Juvenile's song or a mouse in my lab, CharmCitySweetie became her particular adult self. How did she come by her Boston accent, her heterosexuality, her curvy figure, her sense of humor, her chilly feet, her taste for IPAs, and her perfect pitch? It turns out that there's a different explanation for each of these traits.

How we become unique is one of the deepest questions that we can ask. The answers, where they exist, have profound implications, and not just for internet dating. They inform how we think about morality, public policy, faith, health care, education, and the law. For example: If a behavioral trait like aggression has a heritable component, then are people born with a biological predisposition toward it less legally culpable for their violent acts? Another question: If we know that poverty reduces the heritability of a valued human trait like height, should we, as a society, seek to reduce the inequities that impede people from fulfilling their genetic capacity? These are the types of questions where the science of human individuality can inform discussion.

Although investigating the origins of individuality is not just an endeavor for biologists—cultural anthropologists, artists, historians, linguists, literary theorists, philosophers, psychologists, and many others have a seat at this table—many of this topic's most important aspects involve fundamental questions about the development, genetics, and plasticity of the nervous system. The good news is that recent scientific findings are illuminating this question in ways that are exciting and sometimes counterintuitive. The better news is that it doesn't just boil down to the same tiresome nature-versus-nurture debate that has been impeding progress and boring people for years. Genes are built to be modified by experience. That experience is not just the obvious stuff, like how your parents raised you, but more complicated and fascinating things like the diseases you've had (or those that your mother had while she was carrying you in utero), the foods you've eaten, the bacteria that reside in your body, the weather during your early development, and the long reach of culture and technology.

So, let's dig into the science. It can be controversial stuff. Questions about the origins of human individuality speak directly to who we are. They challenge our concepts of nation, gender, and race. They are inherently political and incite strong passions. For over 150 years, from the high colonial era to the present, these arguments have separated the political Right from the Left more clearly than any issue of policy.

Given this fraught backdrop, I'll do my best to play it straight and synthesize the current scientific consensus (where it exists), explain the debates, and point out where the sidewalk of our understanding simply ends. And if you want to keep your internet dating browser window open as you read, rest assured that I won't judge you.

ONE

It Runs in the Family

IN 1952, DIMITRI BELYAEV, A RUSSIAN GENETICIST, had an idea for a creative and audacious experiment. He was interested in the domestication of animals that had been important to human civilization, such as dogs, pigs, horses, sheep, and cattle. Dogs are thought to be the first domesticated species, derived from Eurasian gray wolves by hunter-gatherers over fifteen thousand years ago.[1] Belyaev wanted to know how some wild wolves, which are famously averse to human contact and occasionally aggressive, evolved into the affectionate and loyal companions we know and love. Why, as first described by Charles Darwin, did domesticated mammals often share certain physical characteristics—like rounder, more juvenile-appearing faces, floppier ears, curvier tails, and patches of lighter fur or hide—in contrast to their wild forebears? And why did most wild mammals have a single, brief breeding season every year, but their domesticated counterparts could often breed twice or more per year?

Belyaev believed that the single most important trait selected for in the initial process of domestication was not size or breeding capacity but tameness. He hypothesized that the defining characteristic of all the animal species domesticated by our ancestors was a reduction in aggression toward, and fear of, humans. To test his theory, he went to some of the industrial-scale silver fox farms that had been established for fur production in the Soviet Union and instructed the animal breeders there to select only the tamest foxes, a tiny fraction of the total, and breed them together. He believed that by repeatedly selecting for tameness over many generations, he could ultimately approximate wolf-to-dog domestication and produce a friendly, loyal, doglike fox.

In carrying out these experiments, Belyaev hoped to avoid the fate of his beloved older brother Nikolai, who, in 1937, had been executed by the Soviet government for the crime of performing and publishing genetic experiments. Those were dark days for Soviet biology. Stalin's Communist government, eager to elevate an uneducated "common man" to a position of authority in the scientific leadership, promoted the charlatan Trofim Lysenko to director of the Institute of Genetics at the Soviet Academy of Sciences. Lysenko faked his data to show that wheat and barley seeds that had been frozen before planting yielded larger crops when planted in winter, and that the second generation of seeds derived from those crops also acquired enhanced winter growth. He claimed that this method could double food production in the USSR and feed the masses, an assertion extolled in the state-controlled newspaper *Pravda* as a triumph of Soviet science. His seed-freezing techniques were widely adopted in the country but failed utterly, contributing to periods of widespread starvation. Lysenko rejected genetics, a discipline that had thrived in Russia before his rise to power, because simple genetic experiments could disprove his claims. He called Soviet geneticists "Western saboteurs" and, with Stalin's backing, sought to dismantle the discipline. Those who resisted were fired and even imprisoned. The strongest supporters of genetics, like Nikolai Belyaev and the great Russian plant geneticist Nikolai Vavilov, were executed—Belyaev with a rifle, and Vavilov by slow starvation in a prison cell.

Dimitri Belyaev was fortunate to have some political support for his work. A decorated hero of the Russian Army during World War II, he had presided over improvements in the farming of wild foxes, sable, and mink for fur production. This effort was key to the Soviet economy because it brought in large amounts of foreign currency. Mindful of his brother's fate, Belyaev conducted his domestication experiments on remote fox farms, far from the prying eyes of Moscow—first in the forests of Estonia and later in a distant part of Siberia near the Mongolian border. The cover story was that he was studying fox physiology, not genetics. To oversee the endeavor, Belyaev recruited the young scientist Lyudmila Trut, an expert in animal behavior who had been trained at the elite Moscow State University. He gave her explicit instructions: when selecting foxes for breeding, the only trait to be considered was tameness—not appearance, nor size, nor behavior toward other foxes.

There was no guarantee that this fox domestication plan would work. Nonetheless, it was a reasonable supposition. After all, dogs were domesticated from wolves, which are closely related to foxes. Yet previous attempts to domesticate wild zebras—which are so closely related to horses that the two species can sometimes be interbred (a Shetland pony-zebra cross is called a zony)—had repeatedly failed.[2] The reason appears to be that there is not enough genetic variation underlying the trait of tameness in zebras. You can't effectively pick the tamest zebras for breeding if there

FIGURE 1.
Dr. Lyudmila Trut with one of her domesticated foxes. Used with permission of the BBC. Photo by Dan Child.

aren't any slightly tame zebras to start with. Fortunately, that wasn't the case with Trut and Belyaev's foxes.

When Lyudmila Trut first slowly introduced her hand into the fox cage, she wore a thick padded glove and held a small stick. The most common reaction to this gentle intrusion was snarling and biting. Other foxes cowered, highly agitated, in the rear of the cage. But about 10 percent of the foxes stayed calm throughout, observing her intently but not approaching.[3] These were the animals that she selected for the first round of breeding. Trut was also careful not to breed closely related foxes and thereby introduce inbreeding artifacts that could confound the experiment. To increase the probability that the observed tameness resulted purely from genetic selection, the foxes were not trained and their interactions with humans were strictly limited.

Trut's initial finding, that there was some partial tameness to serve as a basis for subsequent breeding, was encouraging. But the experiment could still easily fail in a different way: it simply might take too many generations to see any significant changes in fox behavior. It has been suggested, from analysis of the archeological record, that wolf-to-dog domestication proceeded in fits and starts, beginning thousands of years ago. Trut and Belyaev didn't have that much time and were limited by the slow pace of fox breeding: one mating season per year. So it was cause for joy when, only four years into the experiment, clear behavioral changes emerged. A few of the fourth-generation foxes showed no aggression or fear, and even displayed doglike tail wagging in response to humans. By the sixth generation, some of the fox pups exhibited whining, licking, and whimpering behavior as they eagerly sought human attention. Today, over 80 percent of the adult foxes derived from these crosses are as loyal and tame as any domesticated dog (figure 1).[4]

If you wish, you can go on the internet and obtain your own tame fox from Trut and Belyaev's experiment, delivered from Siberia to you for $9,000, shipping included.[5] But be aware that, while domesticated foxes are much friendlier than those in the wild, they are much harder to train than dogs. "[You can be] sitting there drinking your cup of coffee and turning your head for a second, and then taking a swig and realizing, 'Yeah, Boris came up here and peed in my coffee cup,'" said domesticated fox expert Amy Bassett. "You can easily train and manage behavioral problems in dogs, but there are a lot of behaviors in foxes . . . that you will never be able to manage."[6]

═══

THE ORIGINAL FARMED SILVER foxes looked like wild foxes: they had erect ears, low-slung tails, and uniformly silver-black fur, save for a white tail tip. As breeding for tameness continued through the generations, the foxes often developed floppy ears, shorter, curved tails, and patchy, pale fur, particularly on the face. They reached sexual maturity earlier than wild foxes, and some even bred twice per year. It is important to emphasize that the only criterion used for breeding was tameness; the other physical traits just came along for the ride. The remarkable thing is that these particular bodily changes have emerged in many other domesticated animals—from cattle to pigs to rabbits—at various times in history.

When Trut and Belyaev measured the levels of resting stress hormones produced by the adrenal glands, they found significant reductions in the tame foxes. They also found that levels of the neurotransmitter serotonin and its metabolites were increased in the brains of the tame foxes, which is consistent with a reduction in aggressive behavior. One overarching hypothesis for the biochemical, behavioral, and structural changes seen in domesticated foxes and other animals is that their development is somehow arrested at an earlier state than their wild cousins. Perhaps the variation in genes responsible for developmental timing is what gives rise to variation in tameness. When animals are bred for tameness, the other youthful traits noticed by Darwin—like floppy ears, round faces, and curly tails—follow along.

TRUT AND BELYAEV SHOWED that a behavioral trait (tameness) in foxes is heritable, that it can be changed by selective breeding in just a few generations, and that physical changes will accompany selection for this trait. Can these conclusions about the heritability of behavioral and physical traits from the fox taming experiment be usefully applied to us? After all, we humans are not confined to cages in Siberia. And, for the most part, we choose our own mates, rather than having them forced upon us by alien overlords. We even have OkCupid and Bumble to expand our mating possibilities.

Insights about the heritability of human traits can be gleaned from studies of twins. This type of analysis can be used to estimate the degree of variation in a trait that is heritable within a particular group of people (or foxes), from 0 to 100 percent. The key thing to remember about heritability is that it measures variation across an entire population, not individuals. Just because a particular trait is 70 percent heritable doesn't mean that, for any individual from that population, genes are responsible for 70 percent and other factors for 30 percent.

Heritability estimates from twin studies may be used for both easily measured physical traits, like height or resting heart rate, and behavioral traits like shyness, generosity, or general intelligence, which are somewhat more subjective and harder to measure. One of the challenges with behavioral traits, which are typically measured by direct observation or with a survey, is that they are culturally constructed. The definition of and necessary criteria for the trait of shyness is probably different in Japan than it is in Italy. Concepts of generosity will not be identical for the city dwellers of Pakistan and the Hadza people of Tanzania. What this means is that the assessment of behavioral traits in individuals will be convolved with cultural factors if the individuals come from different cultural backgrounds, even if they live in the same location.

Here's how heritability estimates work: Fraternal twins are conceived when two eggs are released during the same ovulatory cycle and each is fertilized by a separate sperm cell. The two fertilized eggs then develop separately into two embryos. Fraternal twins are as genetically similar to each other as any other pair of siblings. On average, they share 50 percent of their genes.[7] Since fraternal twin embryos inherit their sex-determining X and Y chromosomes independently, fraternal twins are as likely to be the same sex (boy/boy or girl/girl) as the opposite sex (boy/girl or girl/boy).

By contrast, identical twins arise from a single fertilized egg that then divides to form two embryos early in development. Each twin inherits the same version of each gene from their parents, and so they are genetically identical. Because identical twin embryos also inherit the same arrangement of sex-determining X and Y chromosomes, they are always the same sex. This means that if you see mixed-sex twins, they must be fraternal, not identical.

In one simple twin study design, a particular trait, like height, is measured in members of a large number of fraternal and identical twin pairs. The difference in height is calculated for each twin pair, and then the results are compared between the fraternal and identical groups.[8] One study of this type has shown, for example, that the average height difference between fraternal twins is 4.5 centimeters, whereas it is 1.7 centimeters for identical twins. A crucial assumption in these types of twin studies is that both twins (identical and fraternal) have been raised together, in the same household, at the same time, and will thereby have a highly shared social and physical environment, at least during childhood. Therefore, the smaller average difference between identical twins is attributed to their greater degree of genetic similarity. When these values are plugged into a standard equation, we can estimate the degree of heritability of a trait, which is about 85 percent for adult height, at least in affluent countries where basic nutrition needs are met. One can also estimate the degree of variation in height that is attributable to the twins' shared environment, which is about 5 percent, and to the twins' unshared environment, which is about 10 percent. Those interested in the calculation of these values are invited to check this endnote.[9]

For most twins, the shared environment is dominated by experiences in the family (both social, like being read to, and physical, like the particular foods on the dinner table) but can also include certain shared experiences at school and in the community, as well as the shared exposure to foods and infectious diseases. Unshared environment is a sort of grab bag for all of the other types of random experience, both social and biological, that individuals do not share. Importantly, this estimate of non-shared environment will also include the random nature of both fetal and postnatal development of the brain and body, which we shall explore in chapter 2.[10]

This type of twin analysis can be applied to any trait, not just those that are continuously variable and easily measured, such as height or weight. For example, it can be used to analyze responses to a survey question like "In the last year, have you ever felt sexual attraction to a member of your own sex?" If sexual attraction had no heritable component, we'd expect that the percentage of twin pairs where both answered yes would be roughly the same for identical and fraternal twins. Conversely, if sexual attraction were entirely heritable, then we'd expect that every homosexual/bisexual identical twin would have a homosexual/bisexual twin sibling (and every straight identical twin would have a straight twin sibling). It turns out that the best estimates to date (from a population of 3,826 randomly selected twin pairs in Sweden) are that, in men, about 40 percent of the variation in sexual orientation is heritable with no detectable effect of shared environment and 60 percent is attributable to unshared environment.[11] Forty percent is a significant fraction, but it still leaves room for plenty of other nonheritable factors. We'll discuss the emerging science of sexual orientation and identity in chapter 4.

There have been critiques of these types of twin studies. Some researchers have claimed that studies comparing identical and fraternal twins raised together overestimate the heritable contribution to a trait because family members, friends, and teachers often treat identical twins more similarly than fraternal twins. This could come about in many ways, from the foods they are served to the ways in which people interact with them. Other researchers have claimed the opposite problem: they argue that since identical twins raised together seek to differentiate themselves socially from each other to a greater degree than fraternal twins, such a comparison underestimates the genetic contribution to a trait (particularly a behavioral one). In either case, the key assumption of equal shared environments between identical and fraternal twins would be violated. There have been passionate arguments for and against the validity of studies of twins raised together, and we won't engage in a blow-by-blow recap of those brawls here. My own reading of the literature leads me to believe that, in most cases, the unequal shared environment problem in studies of twins raised together is small and rarely invalidates the general estimates of heritability that result.[12] Nonetheless, it would be best to have a twin study design that would cleanly estimate heritability without the muddled assumption of equal shared environments.

On February 19, 1979, (at which point the tame-fox breeding experiment had been underway in the Soviet Union for over twenty-six years), the local newspaper in Lima, Ohio, reported a fun human-interest story about identical twin brothers who had been adopted by different families and raised completely apart, only to reunite at age thirty-nine. The twins were born in 1939 to a fifteen-year-old unwed mother, who immediately put them up for adoption. They were separated four weeks later, when one was adopted by Ernest and Sarah Springer, who brought him to their home in Piqua, Ohio. The second boy was adopted two weeks later by Jess and Lucille Lewis of Lima, Ohio, a town about forty-five miles away from Piqua. For reasons that have never been explained, both couples were told that their adoptive child had a twin who died at birth.[13]

But when Lucille Lewis was finalizing the legal adoption of her son, by then a toddler, a clerk at the county courthouse let the cat out of the bag. She told her, "They named the other little boy Jim, too." In an interview with *People* magazine, Mrs. Lewis said, "I knew all those years that he had a brother, and I worried whether he had a home, and whether he was all right." She waited until her son turned five before telling him about his twin. Jim Lewis couldn't explain why, at the age of thirty-nine, he finally contacted the court to put him in touch with his brother. The *Lima News* reported that Jim Lewis called Jim Springer, took a deep breath, and asked, "Are you my brother?" At the other end of the line, Jim Springer answered, "Yep." And so, the twins were reunited.[14]

When the Jim twins reunited, they were not mirror duplicates in either appearance (figure 2) or temperament. Nonetheless, a series of striking similarities emerged. Both brothers worked in law enforcement and enjoyed carpentry and drafting as hobbies. On vacations, they liked to drive their Chevrolets to Pass-a-Grille Beach in the Florida panhandle. In school, both had excelled in math and struggled with spelling. Both had married women named Linda, only to divorce and remarry women named Betty. Both had sons: James Alan Lewis and James Allan Springer. And, most tellingly, they preferred to wash their hands both before and after peeing.

It's not surprising that these anecdotes were broadly appealing to readers and that the story of the Jim twins quickly made its way around the world. The day after the first story of their reunion appeared in the *Lima News*, it was reprinted in the *Minneapolis Star Tribune*, where it caught the eye of Meg Keyes, a psychology graduate student at the University of Minnesota. Keyes had recently taken a course with Professor Thomas Bouchard Jr. on individual behavioral differences. When she showed Bouchard the article, he immediately recognized how interesting it would be to study the Jim twins, and soon. He was quoted in the *New York Times* as saying, "[To study the Jim twins] I'm going to beg, borrow and steal and even use some of my own money if I have to. It is important to study them immediately because now that they have gotten together they are, in a sense, contaminating one another."[15]

Bouchard quickly contacted the twins, who agreed to come to the University of Minnesota to spend six days undergoing a battery of psychological and medical tests and interviews. More stories of behavioral and physical similarities emerged. Both crossed their legs in the same way and suffered from chronic headaches and a heart condition. Both were described as "patient, kind, and serious." Both had rapidly gained ten pounds at exactly the same age. These anecdotal similarities were tantalizing, but analysis of a single identical twin pair, even one as striking as the Jim twins, did not allow Bouchard to reach the holy grail: to estimate the heritability of traits without the potential confound of the equal environment assumption. That would require him to compare a sizeable population of identical twins with an equally sizeable population of fraternal twins raised apart.

FIGURE 2. Jim Springer and Jim Lewis pose for a photo shortly after being reunited in 1979. Photo courtesy of Nancy L. Segal and the Jim twins. Used with permission.

When the study of the Jim twins began, Bouchard assumed that they would be a one-off. Other researchers had tried to analyze twins raised apart but had access to so few twin pairs that their results were statistically weak. Bouchard imagined that he would have the same problem, that the cost of finding many twins raised apart would be prohibitive. What he didn't count on was the public's insatiable appetite for Jim twin stories. They appeared in newspapers, magazines, and on all the major television shows of the day. Some newly separated twin pairs emerged after the Jims appeared on *The Tonight Show* with Johnny Carson, others after seeing them on Dinah Shore.

This unprecedented publicity allowed Bouchard to found the Minnesota Study of Twins Reared Apart (MISTRA), which ran for twenty years and analyzed eighty-one identical and fifty-six same-sex fraternal twin pairs.[16] In collaboration with fellow University of Minnesota psychologist David Lykken, the study also compared twins reared apart with twins reared together. MISTRA was a major advance in twin research. The largest and most productive investigation of this type, it produced good estimates of the heritable contribution to variance in many physical traits, like body mass index (about 75 percent) and resting heart rate (about 50 percent), and behavioral traits, like extraversion (about 50 percent) and schizophrenia (about 85 percent).

One main conclusion of MISTRA and related studies was that most human traits, regardless of whether they are physical or behavioral, have a significant heritable component, usually ranging from 30 to 80 percent. Rarely are traits either entirely heritable or entirely nonheritable (we'll talk about some notable exceptions to this later). The other main conclusion is that variation in certain traits, like IQ, is weakly heritable (about 22 percent) when tested at age five but becomes strongly heritable once school is well underway at age twelve (about 70 percent), and then remains so across the lifespan. Correspondingly, the variation in IQ explained by the shared environment is about 55 percent at age five (when most experience has been within the family) but falls to undetectable levels by age twelve, at which time children have been exposed to a broader range of experiences.[17] Those of you who are doing the arithmetic will notice that the variations accounted for by heritability and shared environment are not adding up to 100 percent. That difference is the aforementioned term "unshared environment," which, in addition to unshared social experience, also includes the random process of development. More on this in chapter 2.

For decades, the dominant thinking in the field of psychology, and in society at large, was that the most important determinant of one's adult personality was the influence of immediate family, particularly the parents. This idea came from the twentieth-century psychological movement called behaviorism, which held that humans come into the world as blank slates, ready to be molded by social experience. As a result, it was quite a shock when the MISTRA experiments showed significantly higher correlations in personality measures in identical twin pairs than in fraternal ones. The main result was that about 50 percent of the variation in personality can be accounted for by heritability. This held for all five major standard scales of personality (openness, conscientiousness, extraversion, agreeableness, and neuroticism; abbreviated as OCEAN) and directly contradicted the blank slate hypothesis of the behaviorists.

Most psychologists were guessing that the remaining 50 percent of the variation would be largely explained by social dynamics within the family. By comparing identical twins raised together with identical twins raised apart, the MISTRA researchers estimated the contribution of "shared environment" to individual personality—a factor that includes social experience in the family as well as things like shared nutrition and shared exposure to communicable diseases. To the psychologists' surprise, shared environment made little or no contribution to variation in personality measures (typically less than 10 percent). It's not just identical twin results that support the idea that shared environment plays a tiny role in explaining individual personalities. Fraternal twins who grow up together are no more similar in personality than those raised in different families, and unrelated adoptive siblings raised in the same family are barely alike at all.

The failure of shared environment to affect personality goes against some popular ideas about the influence of parents. But these twin study results don't say that parental behavior is unimportant. Rather, they show that, beyond some minimum level of parental support and encouragement, extra attention doesn't produce large effects on personality as measured by questionnaires administered in the lab.

Importantly, personality is not the totality of one's character. Parents can inculcate work habits and teach specific skills, like weaving or car repair. And they can transmit philosophical, religious, or political opinions that are not measured by the OCEAN personality tests. For example, altruism, sharing, and other prosocial behaviors appear to be influenced by shared environments to a greater degree than other behavioral traits.[18] Religiousness is another trait where significant variation is contributed by both heritable factors and shared environment. Importantly, while one's likelihood of having religious beliefs is influenced by both heredity and shared environment, the specific religion you choose has no hereditary component. Your genes might contribute to making you religious, but they will not specify a Hindu or Wiccan or Roman Catholic faith—that's mostly a family and community affair.

Another well-entrenched idea about the influence of family on personality has to do with birth order. First children are generally thought to be socially dominant, less fearful, and more novelty seeking and risk-taking as compared to their later-born siblings. And if one observes children at home, this stereotype is borne out. Parents treat firstborns differently than their later children, and firstborns both care for and boss around their younger siblings throughout childhood. Indeed, these social patterns often persist within the family as the children become adults. But remarkably, study after study has failed to find that the domineering qualities of firstborns are present outside of the family.[19] Neither at school, nor on sports teams, nor in the workplace do firstborn children show an unusual degree of social dominance or any other personality trait. In retrospect, this makes sense. The firstborn child who is the oldest and biggest at home no longer enjoys that same status on the playground, in the classroom, or in other locations outside of the family.

IF THE JIM TWINS had not been so alike and generated such appealing stories and media attention, the MISTRA study might not have happened at all. The Jim twins were certainly among the most similar in the study and hence not the most representative example of identical twins raised apart. Bouchard noted this issue: "There probably are genetic influences on almost all facets of human behavior, but the emphasis on the idiosyncratic characteristics is misleading. On average, identical twins raised separately are about 50 percent similar [in behavioral measures]—and that defeats the widespread belief that identical twins are carbon copies. Obviously, they are not. Each is a unique individual in his or her own right."

When the MISTRA results first began to be published in the 1980s, the reception was not entirely positive. The evidence that there was a strong heritable component to complex behavioral traits like novelty seeking, traditionalism, and general intelligence, while embraced by some, was met with skepticism and hostility by others, particularly the adherents of behaviorism. Bouchard and his coworkers were called frauds, racists, and Nazis. Some opponents sought to have him fired from the University of Minnesota. However, over time, the MISTRA findings, on both behavioral and physical traits, were replicated by several well-controlled studies of twins raised apart. An important caveat is that, to date, most of these studies have been performed among more affluent populations in countries like Japan, the United States, Sweden, and Finland, where nutritious food, medical care, and decent schools are widely available. While there are still arguments to be had, most biologists today accept that most behavioral and physical traits have a substantial heritable component.[20]

Danielle Reed, a scientist from the Monell Chemical Senses Center, credits Bouchard's work with expanding our understanding of heredity. "He was the trailblazer," she says. "We forget that 50 years ago things like alcoholism and heart disease were thought to be caused entirely by lifestyle. Schizophrenia was thought to be due to poor mothering. Twin studies have allowed us to be more reflective about what people are actually born with and what's caused by experience."[21]

FOR YEARS, PEOPLE HAVE argued about the origin of human traits. The most politically and emotionally fraught of these arguments concerns IQ tests as a measure of intelligence. Can intelligence be determined by heredity, environment, or something else? Are IQ tests even valid cross-culturally? The results from MISTRA and certain other twin studies have estimated that about 70 percent of the variation in IQ test scores is heritable. The first and most obvious point is that 70 percent is not 100 percent—this value still leaves room for significant environmental influences. The second point is more subtle. Estimates of heritability are only valid for the population that is analyzed. While the MISTRA investigators did not seek out a particular type of twin pair for their study, its population was overwhelmingly white, midwestern, and middle class, and so the 70 percent heritability estimate does not necessarily apply to other populations.

Perhaps it's easier to think about heritability for human populations using a less politically sensitive trait, like height. In affluent populations, with good access to nutritious food, clean water, decent sleep, and basic medical care, about

85 percent of the variation in height is heritable. But if we look at a population that does not have these advantages, like poor people in rural India or Bolivia, then only about 50 percent is heritable. Without access to basic nutrition (including sufficient protein) and treatment for diseases (mostly infectious ones), poor people are not able to reach their genetic potential for height.[22] Stated another way, the heritable and environmental components of a trait are not simply summed up. Heredity interacts with the environment, providing the potential for a trait, but environmental conditions influence whether that potential will be fully realized.

It's the same situation for IQ test scores: children without access to basic human needs—not just nutrition, health care, and sanitation, but decent schools, books, sufficient sleep, and the freedom to explore and be curious—cannot fulfill their genetic potential for general intelligence. Crucially, the degree of variation in general intelligence explained by heredity is lower for poor populations than for those whose basic needs have been met.[23] To me, the political and moral lesson from the study of trait heritability is clear: if you want to improve the lives of humanity as a whole, the first job is to make sure that everyone has his or her basic needs met in order to fulfill her or his genetic potential for positive human traits. We'll return to this issue when we explore population differences and concepts of race and racism in chapter 8.

TWIN STUDIES CAN MEASURE the average contribution of heritability to variation in human traits across a population, but they cannot reveal the underlying biological mechanisms responsible for this variation. In order to do that, we'll need to consider the biochemical machinery of life. Heredity is encoded in DNA, which resides in the nucleus of cells. It is organized into genes, each of which contains the instructions to make a different protein. Some proteins are structural: they are the girders and cables that determine the shapes of cells. Others have specific biochemical functions, such as creating or breaking down an important chemical in the body, like a digestive enzyme in the stomach. Yet other proteins are receptors, specialized micromachines that allow cells to respond to chemical signals like hormones or neurotransmitters. Still more are transducers that help us sense things about the world around us, like the proteins in the retina that allow us to see light or those in the inner ear that allow us to hear sounds, converting those forms of energy into electrical signals that ultimately travel to the brain.

DNA is composed of long chains of chemical groups called nucleotides that come in four flavors: A, C, T, or G. Humans have about three billion nucleotides organized into about nineteen thousand different genes, with vast gaps of more poorly understood DNA between them.[24] Together, all of this DNA is the human genome. We now know the complete nucleotide sequence of the human genome, as well as that of some plants, animals, and bacteria. In turns out that nineteen thousand genes is not an unusual number for an animal. The tiny roundworm *C. elegans* has about the same number. By comparison, a fruit fly has about thirteen thousand and a rice plant about thirty-two thousand. A particular type of poplar tree is the present winner with about forty-five thousand genes. Clearly, the number of genes in an organism's genome does not determine the anatomical complexity, much less the mental capacity of that critter or plant.[25]

On average, considering the entire DNA sequence (both the genes and the stretches of other DNA between them), each human is about 99.8 percent similar to any other human, 98 percent similar to a chimpanzee, and 50 percent similar to a fruit fly. This is because, if you go back far enough in evolutionary time, about eight hundred million years, humans, chimps, and fruit flies all share a common ancestor.

If only a 2 percent difference separates us from chimps, then it follows that small differences in the DNA sequence can sometimes have a big effect on traits. Indeed, there are certain locations in the human genome where a change in a single nucleotide (called a point mutation) will be fatal. Sometimes, if the change acts early in development, the embryo will die. There are other places where changes in a single nucleotide can cause a serious disease. For example, certain tiny changes in the gene that instructs the production of an enzyme that metabolizes the amino acid phenylalanine will break it. As a result, when an infant carrying this mutation eats phenylalanine-containing foods, the amino acid builds up to toxic levels and impairs development of the brain and other organs, producing the disease phenylketonuria (known as PKU).[26] There are many other examples of single-nucleotide mutations in genes, but it's worthwhile to note that, unlike PKU, most have no functional consequences at all.[27]

We generally carry two copies of each gene, each called an allele: one from our mother and one from our father. For most genes, both the maternal and paternal copies are active.[28] So, to have PKU, you need to inherit broken copies of the gene instructing production of phenylalanine from both your mother and your father. This qualifies PKU as a recessive genetic disease. There are other genetic diseases that are inherited in a dominant fashion, like Marfan syndrome (a disease of overly stretchy connective tissue), where receiving a single copy of a particular gene variant from either parent is sufficient to produce the disease.

═

HERE'S A FUN FACT you can use to impress your friends: everyone has either wet or dry earwax. If your ancestors are from Europe or Africa, there's a very high chance (greater than 90 percent) that you have the wet type. If your ancestors are from Korea, Japan, or northern China, then you almost certainly have the dry type. If your people hail from South Asia, or if you are of mixed northeast Asian and European/African ancestry, then your chance of having dry earwax is somewhere in the middle. To study the genetics of earwax, a group of scientists led by Norio Niikawa of the Nagasaki University School of Medicine obtained DNA and earwax samples from people all around the world.[29]

They determined that the dry earwax trait is due to a single-nucleotide mutation in a gene that controls various forms of secretion (*ABCC11*). Like PKU, having dry earwax is recessive; it requires inheriting a mutant form of the gene from both of your parents. To put this back in the context of twin studies, the dry earwax trait (and the PKU trait) is 100 percent heritable. There is no contribution of either shared or individual environment. It doesn't matter how your parents raised you or what kind of experiences you had in school or what foods you ate. If you inherited two mutant copies of the dry earwax gene variant, you are going to have dry earwax—end of story.

The mutation in the *ABCC11* gene that causes dry earwax also eliminates armpit odor.[30] That's the main reason why the subway at rush hour in Seoul smells so much better than it does in New York City. The *ABCC11* gene plays a role in secretions from the apocrine glands, the special sweat glands in the armpits (and the external genitals) that secrete oily substances that are then metabolized by bacteria to create funky odors.[31] Because of the *ABCC11* mutation, nearly all Korean people (and most Japanese and northern Han

Chinese people) have odorless armpits along with dry earwax. It is rumored that, in some cases, armpit odor has been a sufficient condition to excuse Japanese men from military service. Stinky armpits are so rare in Japan that some Japanese people who have the stinky-armpit trait seek surgical removal of their armpit apocrine glands. But anxiety about armpit odor is not just a Japanese phenomenon. Revealing the power of advertising and social conformity, one study showed that, among those rare women in the United Kingdom who have the odorless-armpit trait, most (78 percent) still buy and use deodorant.[32]

═══

After hearing about PKU and dry earwax, one could begin to think that there are single genes that drive many human traits. In fact, such traits are quite rare, occupying the far end of the heritability spectrum. The other end consists of traits, like speech accent, that appear to have no heritable basis whatsoever. While there are heritable factors that contribute to the quality of your voice (high or low pitched, resonant or thin, raspy or clear), and these voice qualities will be equally evident in both your speech and singing voice, your accent is entirely determined by your experience of hearing the speech of others. There is no genetic contribution at all. Interestingly, the speech that we imitate most strongly is that of our peers, not our parents. That is why the children of immigrants tend to have the accents of the community where they were raised.

Most traits are neither entirely heritable, like earwax type, nor entirely environmental, like speech accent. Rather, 30 to 80 percent of their variation across a population can be explained by genes. In recent years, a new approach, called a genome-wide association study (GWAS), has helped to show why that is the case. Let's say that you want to know what genes contribute to variation in height (which we know is about 85 percent heritable in affluent populations). You'd assemble thousands of people chosen at random, spanning the range of adult human height. Then you'd collect DNA samples and look at variation across all nineteen thousand or so genes in the genome, as well as the long stretches of DNA between genes. In fact, this very study was done with over seven hundred thousand people, and it showed that height was not determined by changes in a single gene or even a handful of genes, but rather by variation in at least seven hundred genes. Some of these genes were known to contribute to the growth of bone, muscle, or cartilage and so weren't a surprise. But many others would never have been guessed beforehand, reflecting the fact that there are many genes in the genome whose functions remain poorly understood.[33]

There is no single height gene. Rather, there are many genes, and variation in each contributes a small amount to overall height (and each of these genes also influences traits other than height). In addition, the variation in each of these many genes does not merely sum up, but rather combines in sometimes complex and unpredictable ways. Variation in two different genes can add up to more than the sum of their small effects; 1 + 1 = 5, if you will. Other times, two genes can cancel each other out, yielding a 1 + 1 = 0 situation.

The same is true of behavioral traits. There is no single gene for religiosity, neuroticism, or empathy. Genes contain the information to instruct the production of proteins (like the D2-type dopamine receptor or the enzyme tyrosine hydroxylase), not behavioral traits like shyness or risk-taking. A disorder like schizophrenia or a structural trait like height can be highly heritable (both about 85 percent) but also determined by the concerted interaction of *many hundreds of genes.* Please remember this the next time you see a news report about "the IQ gene" or "the empathy gene" or some such nonsense.[34]

WITH THIS BACKGROUND ON trait heritability and genes, let's return to Trut and Belyaev's fox-taming experiment. One way to discover which genes are involved in the emerging trait of tameness would be to take a page from the human height studies: do a GWAS by taking DNA samples from many tame and wild foxes and compare variation across the genome with scores of tameness. Another way is called a candidate gene approach. Recent work from Monique Udell and her coworkers at Oregon State University has shown that variation in two adjacent genes in dogs is strongly associated with tameness and extreme friendliness. Deletion of these same genes (and other nearby genes) occurs in some humans and causes Williams-Beuren syndrome, one symptom of which is extreme friendliness. These findings have led to the interesting hypothesis that one important event in dog domestication has been changes in these two genes that mimic aspects of human Williams-Beuren syndrome.[35] Soon, we will know if the Siberian tame foxes have similar mutations in these two genes, which would be a big step in understanding the emergence of tameness in particular, and novel behavioral traits more generally.

TWO

Are You Experienced?

THE PROBLEM, I'M CONVINCED, IS THAT IT FALLS trippingly off the tongue: "Nature versus nurture." With that great alliteration, rhyme, and snare drum beat, it's just fun to say, like "Might makes right" or "If the glove doesn't fit, you must acquit." The British polymath Francis Galton didn't invent this catchy expression,[1] but he popularized it starting in 1869, and it's been messing things up ever since. First of all, why say "nature" to mean "heredity"? The word nature typically means the entirety of the natural world, as in "the wonders of nature," or the essence or moral character of something, as in "the better angels of our nature." But it never means heredity, except in this one idiosyncratic phrase.

Then there's "versus." The idea that nature and nurture must be in opposition to explain human traits is silly. We know (although Galton and his contemporaries did not) that a few traits (like earwax type) are entirely hereditary, while others (like speech accent) are entirely nonhereditary, but that most traits fall somewhere in between. Even more crucially, we know that nature and nurture interact in various ways to determine traits: To have the symptoms of PKU, you need to both inherit two broken copies of the relevant gene and eat foods rich in phenylalanine. Similarly, you won't be able to reach your full genetic potential for height if you are malnourished or chronically infected. If you're born with athletic talent, you're more likely to seek out opportunities to play sports and improve with practice. The oppositional construction of "nature versus nurture" is just wrong.

But the part of this horrid expression that really chaps my ass is "nurture." The word means how your parents raised you—how they cared for and protected you (or failed to do so) when you were a child. But, of course, that's only one small part of the nonhereditary determination of traits. As we will explore in this chapter, a more correct term would be "experience," which I mean in the broadest sense. Not just social experience and not just the experience of events that you have stored as memories, but rather every single factor that impinges upon you, from the moment that the sperm fertilizes the egg to your last breath. These experiences start even before the embryo implants in the womb and encompass everything from the foods your mother ate while she was carrying you in utero to the waves of stress hormones you secreted on the first day of your first real job.

And there's another important factor that is neither heredity nor experience. That's the random nature of development, particularly the self-assembly of the brain and its five hundred trillion connections. As I mentioned earlier, developmental randomness is a large part of what we measure in twin studies in the category of non-shared environment. This self-assembly is guided by the genome, but it is not precisely specified at the finest levels of anatomy and function.

The genome is not a detailed cell-by-cell blueprint for the development of the body and brain, but rather a vague recipe jotted down on the back on an envelope. The genome doesn't say, "Hey you, glutamate-using neuron #12,345,763! Grow your axon in the dorsal direction for 123 microns and then make a sharp left turn to cross to the other side of the brain." Rather, the instruction is more like, "Hey, you bunch of glutamate-using neurons over there! Grow your axons in the dorsal direction for a bit and then about 50 percent of you make a sharp left to cross the midline to reach the other side of the brain. The rest of you, turn your axons to the right." The key point is that the genetic instructions for development are not precise. In one growing identical twin, 40 percent of the axons in this area will make the left turn; in an another, 60 percent will. The example here is from the brain, but the principle applies to all the organs. That's the main reason why identical twins, who share the same DNA sequence and nearly the same uterine environment, are not born with wholly identical bodies, brains, or temperaments.

This means that your individuality is not a matter of "nature versus nurture" but rather "heredity interacting with experience, filtered through the inherent randomness of development." It's not nearly as fun to say but, unlike the former expression, it's true. The exciting part is that we now have a general understanding of the molecular mechanisms by which heredity, experience, and developmental randomness interact to make you unique. Let me tell you about them.

NEARLY EVERY CELL IN your body contains your entire genome—all nineteen thousand or so genes and the vast stretches of DNA in between them.[2] Yet, in a given cell, only some of these genes will ever be activated to instruct the production of a protein, a process called gene expression. When you think about it, this makes sense. You don't want the cells that form the hair follicles in your scalp to be turning on the genes to make insulin, and you don't want the cells of your pancreas to be growing hair. For example, most of the electrically excitable cells of the nervous system, the neurons, express about thirteen thousand genes. Of those, about seven thousand perform general cellular housekeeping functions, and so most other cells in the body express them too. There are about four hundred genes that tend to be expressed in much higher levels in neurons than in other cell types. Some specialized genes are shared between tissues. For example, both neurons and heart muscle cells are electrically active, so they share expression of certain genes that are required to generate electrical activity.

Doing the arithmetic, we can calculate that there are about six thousand genes that are never expressed in neurons.[3] There are several ways to shut genes off so that they cannot be used to instruct the production of proteins. The longest-lasting way involves attaching small, globular chemical structures called methyl groups (—CH_3) along the length of the gene's DNA sequence.[4] That blocks the information in the gene from being read out. The genes that are never expressed in a given cell type are usually shut off by this methylation of DNA.

In addition to those genes that are always shut off in a particular cell type, there are others that might be turned on or off at various times. For example, during childhood, certain growth-related genes are activated in tissues like muscle, bone, and cartilage, but those genes are turned off once a child stops growing. Other genes are turned on or off on a more rapid time scale. There are many genes in neurons and other tissues that turn on every night and are shut off during the day (or vice versa) and still more that are activated within minutes in response to a particular pattern of electrical activity in the nervous system or rising levels of a hormone.

These transient cycles of gene expression are controlled by different mechanisms. One involves modifying ball-shaped proteins called histones, around which the strands of DNA are wound. Attaching various chemical groups to histones can allow the DNA to unwind, which is a first, necessary step for gene expression. Other chemical groups can prevent this unwinding and therefore block gene expression. Another regulatory step involves proteins called transcription factors, which bind to a section of DNA near the start site of a gene and, in so doing, turn on expression of that gene. In many cases, genes need several transcription factors, all working together, for expression to start and the protein to be made.[5] The regulation of gene expression—by the action of transcription factors, or the attaching of various chemical groups to DNA or histone proteins—is epigenetics. The crucial point here is that none of these mechanisms alter the underlying sequence of As, Cs, Ts, and Gs. That's why it's called epigenetics, rather than genetics.

Gene expression is exquisitely regulated. Genes can be turned on and off in different cell types, at different times, in response to all forms of experience—from hormonal fluctuations to infection to electrical activity from the sense organs. The regulation of gene expression, over both the short and long term, is the crucial place where genes and experience interact to forge human individuality.[6]

IN DECEMBER 1941, THE Imperial Japanese Army invaded the tropics, rapidly overrunning opposition in the steamy colonial enclaves of British Malaya and Burma, Dutch Indonesia, French Indochina, and the American Philippines, as well as the Kingdom of Thailand. Those were heady days for the Japanese military, as they routed, among others, the vaunted British Army. The Japanese enjoyed rapid and decisive military victories, and by March 1942 they stood at the frontier of India. However, not all was well among this tropical fighting force. One serious problem was that many Japanese soldiers were succumbing to heatstroke, rendering them temporarily unable to fight. When army doctors investigated, they found that soldiers from the colder northern Japanese island of Hokkaido had a much higher incidence of heatstroke then their comrades in arms from the subtropical southern island of Kyushu. The reason was that the northern soldiers sweated less and so had reduced evaporative cooling, resulting in dangerously elevated core body temperatures in hot climates. Skin biopsies revealed that northern and southern soldiers had the same total number of sweat glands. These are the eccrine sweat glands, which cover most of the body and secrete saltwater—not the apocrine sweat glands of the armpits and crotch that secrete oily, protein-laden sweat, which were discussed earlier in relation to the dry earwax gene, *ABCC11.* Upon more detailed inspection, doctors discovered that the southern soldiers had more eccrine sweat glands that received nerve fibers carrying sweat-activating electrical signals from the temperature-regulating region of the brain. These are the sweat glands that matter most for keeping your body's core cool on a hot day.

The classic genetic explanation for how this difference came about would be that, over many generations, people living in Kyushu developed differences in their genes compared to those in Hokkaido. These genetic differences would give rise to more innervated sweat glands and better tolerance of hot climates and would be passed down to the offspring of

Kyushu parents. If that were true, then you would imagine that children who were born in Hokkaido but whose parents were from established Kyushu families would inherit Kyushu-typical gene variants and so have larger numbers of activated sweat glands. And, conversely, you'd expect that Kyushu-born and -raised babies of parents from long-established Hokkaido families would have fewer activated sweat glands.

That explanation turned out to be utterly wrong. Instead, the degree of sweat gland innervation is determined by the ambient temperature experienced in your first year and is then locked in for the rest of your life. If you're born in a cold place and you move to a hot place later in life, you're just out of luck—you'll carry your cold-appropriate, reduced-sweating skin with you. However, if you stay in the tropics and have and raise children there, they will have more activated sweat glands and improved thermal regulation.[7]

This potential mismatch between adaptation to the environment of early life and the experience of a different environment in later life seems like a problem for people who move from one place to another, but it may actually be beneficial. Genetic changes in response to the environment are often slow, requiring many generations to emerge. But adaptations determined by early life experience can appear in the very same generation. You and your mate, as northern-born people, may be prone to heatstroke after moving to the tropics; but your child, who carries northern genes, will sweat more extensively and fare better in the heat. This kind of experience-driven developmental plasticity may be part of what has allowed humans to rapidly migrate over long distances. For example, after the first humans crossed the land bridge from Siberia to Alaska, some of them settled all the way down to the tip of South America, spanning many climate zones, within less than one thousand years.

The story of the sweating Japanese soldiers shows us that we can be influenced by early life experiences, like temperature, that are not social in nature. In fact, those experiences can even start in the womb and, in other animals, can be quite dramatic. Some reptiles and amphibians, for example, have temperature-dependent sex determination. Males and females have identical chromosomes, but the pattern of gene expression that determines sex is set by the temperature of the incubating egg during the middle third of development, when the gonads differentiate.[8] When American alligators lay their eggs, those embryos that experience a middle range of temperatures (from 32 to 34 degrees Celsius) will become male, while those either above or below this range will be female. It's not clear if the adult female alligator, when burying her clutch of eggs, is choosing a nest location to influence the sex of her offspring—or if she will be able to modify that choice to keep her offspring from becoming all female as the climate warms.

This process, by which the external physical environment influences the traits of developing animals, can be found in mammals as well. It may sound suspiciously like a validation of astrology, but there's good evidence from several mammals that birth season can influence development. For example, meadow voles born in the fall arrive with thicker fur than those born in the spring, even if both litters have the same parents. This trait is not influenced by the ambient temperature, which is similar in fall and spring. Instead, coat thickness is determined by the changes in day length experienced by the pregnant mother. When meadow voles are brought to the lab, day length can be manipulated with artificial lights. Mothers who experience lengthening days over the course of their twenty-one-day pregnancy, thereby mimicking springtime, give birth to pups with thinner fur. When these same mothers are bred to the same males and subjected to artificial shortening days over the course of their next pregnancy, simulating fall, they give birth to pups with thicker fur.[9]

There are some tantalizing hints from epidemiological studies that birth-season effects are present in humans too. Nicholas Tatonetti and his coworkers at Columbia University analyzed a huge data set: the medical records of over 1.7 million people treated at NewYork-Presbyterian/Columbia University Medical Center who were born between 1900 and 2000. They were looking for statistical associations between a patient's birth month and lifetime incidence of 1,688 different medical conditions, covering the breadth of medicine from middle-ear infections to schizophrenia. Of those 1,688 conditions, only 55 were significantly influenced by birth month, including acute bronchiolitis, which is more prevalent among fall births, and angina (cardiac chest pain) which is overrepresented in those born in early spring (figure 3).[10]

There were some nice things about the design of this study. First, there were no decisions made about which conditions to test or report (which can lead to a bias for reporting positive associations and ignoring negative ones). Second, the population of patients in the database was quite diverse in terms of ancestry and affluence, so the statistics don't just apply to affluent white people, who have been historically overrepresented in the pool of biomedical research subjects. However, there are some important limitations as well. The most obvious is that the patients were drawn from the New York City area, with its particular seasons, range of foods, weather, types of pollution, etc.

More importantly, birth month can reflect different types of influence, both prenatal and postnatal. For example, babies born in the late spring were carried in the later stages of pregnancy during the winter and spring months, when sunlight-driven vitamin D production is weakest. Low maternal

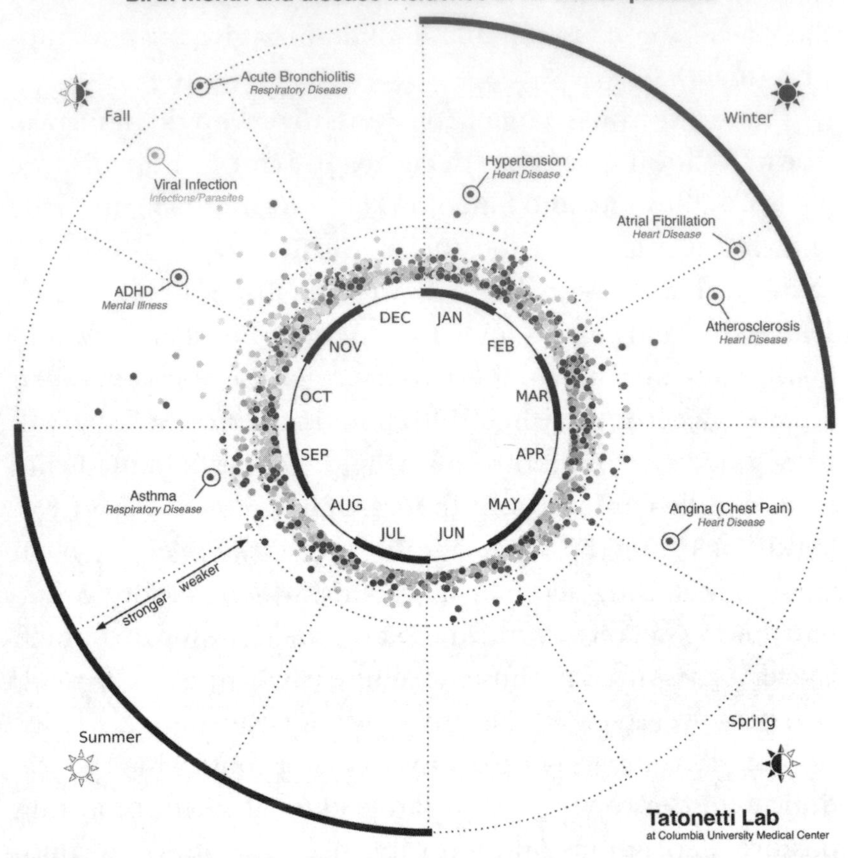

FIGURE 3. Some diseases are significantly more prevalent in people born in particular seasons. In this polar plot, greater distance from the center indicates a stronger statistical association between the disease incidence and birth month. For example, ADHD and acute bronchiolitis are more prevalent in people born in the fall, while atrial fibrillation (a heart problem) is found at higher rates in those born in the winter. This graph applies to the temperate latitudes of the Northern Hemisphere. Figure by Dr. Nicholas Tatonetti. Used with permission.

vitamin D is thought to be a risk factor for certain autoimmune diseases, such as rheumatoid arthritis and systemic lupus.[11] Babies born in the summer and fall arrived into peak indoor dust mite season, which has been suggested to underlie their higher rates of asthma and rhinitis as adults.

And of course, some infectious diseases, like influenza, have varying seasonal incidence.

In addition to physical effects, birth month can also have social influences based on the cutoff date for entering school. If the cutoff day is October 1, then children born in October or November will be among the eldest in their school year, and those born in August and September will be among the youngest. Relative age in school will tend to give a child an advantage in sports. This can also affect medical conditions, as kids who participate in sports tend to have more injuries. Conversely, children younger than their peers are more likely to experience bullying, which can then impact neurological development.

To study the potential effects of age relative to peers, Tatonetti, together with an international cast of collaborators, compiled medical data from 10.5 million patients across six locations in three countries (Taiwan, South Korea, and the United States) with varying latitudes (and therefore seasons), local weather, customs, and school cutoff dates. They calculated the incidence of 133 diseases, chosen so that there would be at least 1,000 patients with that disease at each of the six locations. Of those 133 diseases, only one showed a positive association with age relative to peers: attention deficit hyperactivity disorder (ADHD). Children who are younger relative to their peers in school had an 18 percent higher risk.[12] Why? We don't know. Maybe bullying is a risk factor for ADHD. Maybe it's something else, social or biological. This uncertainty shows us an inherent limitation: epidemiological studies, no matter how carefully designed, cannot prove causality; they can only point us in interesting and useful directions. To go further, we'll need experiments.

THE INFLUENZA PANDEMIC OF 1918 was the deadliest mass infection in modern history. The H1N1 flu strain originated in birds, moved to pigs, and then to humans. The first cases, in the spring of 1918, were reported at Fort Riley, a huge army base in Kansas. The virus spread east through the United States, leaving death and panic in its wake, before hopping the Atlantic to Europe and then Asia during the final months of World War I. The countries fighting on both sides of the war had strong press censorship, which suppressed reporting of the pandemic. Spain, which was neutral, had no such restrictions, and so the Spanish press spread the word. This is why the 1918 strain became known as the Spanish flu, even though it probably originated in North America.[13]

The 1918 pandemic flu was unusual: it had a high mortality rate, typically from secondary bacterial infections like pneumonia, and it was particularly fatal for those in the prime of life (people older than 40 probably had some degree of immunity from exposure to a milder, related flu strain that appeared in 1889). Worldwide, about one in three people were infected and over fifty million people died, including about 675,000 in the United States. To put this in perspective, more US soldiers died from the flu than from combat in World War I. The 1918 flu killed more people in twenty-four weeks than AIDS did in its first twenty-four years in North America.[14]

Many women were pregnant during the 1918 flu season, and about one third of them were infected but survived and gave birth in 1919. The echoes of this pandemic are seen in their children. Caleb Finch of the University of Southern California examined the medical records of soldiers who enlisted in World War II in 1941 and 1942. This sample included 2.7 million men who were born between 1915 and 1922. His team found that men whose mothers had carried them through the time of the 1918 pandemic flu were, on average, about one millimeter shorter than one would expect from comparing them to their fellow soldiers born just before the onset or conceived just after the end of the 1918 flu season.[15] Now, one millimeter of height seems trivial, but in a huge sample like that it's highly statistically significant.

Height is just the tip of the iceberg. The babies born in 1919 grew up to have higher rates of cardiovascular disease (about 20 percent more), perform slightly worse on standardized cognitive tests, and even earn somewhat less money. Perhaps most striking is that the incidence of schizophrenia in this population increased from about 1 percent to about 4 percent. Subsequent studies of other populations with maternal viral exposure in utero have confirmed a similar increased rate of schizophrenia,[16] and have extended it to encompass a higher rate of autism as well.[17]

There are at least two different potential explanations for these findings. One hypothesis is that gene variants that allow the mother (or perhaps the fetus) to survive an influenza infection have other effects, including a reduction in average height and a higher incidence of heart disease, schizophrenia, and autism. Alternatively, we know that viral infection produces activation of the immune system, so perhaps virus-fighting immune cells from the maternal bloodstream or the chemical signals they secrete cross the placenta, enter the umbilical cord, and then affect development of the brain and other organs of the fetus.

If you imagine a young, brilliant, scientist power couple, you may well see Gloria Choi and Jun Huh, of MIT and Harvard Medical School, respectively. He's an immunologist and she's a neuroscientist. In the evening, when their children are put to bed and the dinner dishes are clean, sometimes they talk shop. Choi and Huh had read the scientific literature showing a higher rate of autism in those whose mothers had fought off a viral infection during pregnancy. And they had read a report from Paul Patterson's lab at Caltech showing that, in mice, maternal infection can produce autism-like behaviors, and this process could be blocked by interfering with the action of an immune system signaling molecule called IL-6 in the mother.[18] IL-6 is well known to trigger the production of yet another immune signaling molecule, IL-17a, which can pass from the mother into the developing fetus.

So, Choi and Huh thought that by performing experiments in mice to measure and manipulate IL-17a, they might reveal how maternal infection could change the brain of the developing fetus to produce behaviors associated with autism. To mimic viral infection, they used a well-established method. They injected pregnant mice with synthetic double-stranded RNA about midway through pregnancy, at a time when the neocortex is forming. Then they waited for the pups to be born and grow into adults before subjecting them to analysis. There were two exciting findings. First, the outermost portion of the neocortex was malformed in the mice whose mothers had been infected. Normally, the neocortex looks like a cake with six layers of varying thickness. Now, at various locations in the late-fetal brain, these regular layers were disrupted by protrusions, where blobs of neurons stuck out. When the fetuses from infected mothers grew to adulthood, a different pattern of cortical disruption emerged, which altered local electrical activity and was concentrated in a region called S1DZ. Second, the mice displayed behaviors roughly consistent with autism, including social interaction deficits and repetitive, compulsive behavior (in mice, this behavior is compulsive marble burying). Importantly, when the mothers were infected a few days later in pregnancy, by which time the layered structure of the neocortex had been established, neither the disrupted brain structure nor the autism-like behaviors were produced.

The next step in this inquiry is to understand the cellular and molecular steps that link maternal infection to altered brain development. I apologize in advance for bombarding you with a bunch of names for biomolecules. It's not important that you memorize them. The crucial point is that there is a detailed, specific, and testable hypothesis for maternal-infection-triggered autism here, not just a bunch of hand-waving generalities.

The double-stranded RNA injected into the mother can't cross the placenta into the fetal mice, but it can trigger immune cells in the mother's body, called dendritic cells, to secrete signaling molecules called pro-inflammatory cytokines (please see figure 4 to follow along). These molecules (which have mind-numbing names like IL-6, IL-1beta, and IL-23) stimulate yet another type of immune cell (T helper 17 cell) to secrete IL-17a, a cytokine previously found to be elevated in the blood of autistic kids. IL-17a produced in the mother's body crosses the placenta, flows through the umbilical cord, and binds receptors for IL-17a on developing neurons in the fetal neocortex. Crucially, when molecular or genetic tricks were used to interfere with maternal IL-17a production or signaling, the ability of maternal infection to produce disrupted cortical patches and autism-like behaviors in the pups was abolished. And, to put the icing on the cake, when IL-17a was injected directly into the developing fetal brain, this also produced both neocortical malformation and autism-like behaviors when the pups grew up.[19]

Presumably, when maternal IL-17a binds receptors on the developing fetal brain cells, it causes changes in gene expression in those cells, and those changes lead to the emergence of cortical patches and autism-like behavior. These results are consistent with reports in humans showing that, when autopsies are performed on autistic adults, malformations are sometimes found in the neocortex and that IL-17a is present at higher levels in the blood of some autistic children.[20]

These findings are super-exciting. They describe a molecular pathway to potentially explain the well-established link between maternal infection and the increased incidence of autism. And they point to potential therapies—perhaps interfering with IL-17a or the changes it evokes in the fetal brain could prevent autism from maternal infection. So, it was interesting when the basic finding could not be replicated in Choi and Huh's labs using genetically identical lab mice that came from a different breeder. In mice supplied by Jackson Laboratory, maternal infection produced no elevation in IL-17a and no cortical patches or autism-like behaviors in the adult offspring. Then it was noticed that the original mice, which came from Taconic Biosciences, had a common, innocuous type of bacteria in their gut (called segmented filamentous bacteria, or SFB), while the Jackson mice did not. Sure enough, when Taconic mice were treated with an antibiotic to wipe out SFB, the maternal infection autism effect was abolished, and when Jackson mice had SFB introduced into their guts, the effect was restored.[21]

It turns out that SFB, through a process that is not yet understood, allows T helper 17 cells to differentiate and thereby become competent to secrete IL-17a. The essential point here is that, in order to produce the surge of IL-17a that causes fetal brain trouble, several things all have to happen: the female mouse must be pregnant, she must be carrying the right bacteria in her gut, and she must become infected with a virus. To produce autism in the pups, all of this must happen just as the fetal neocortex is developing, around day twelve of mouse pregnancy. If it happens a bit

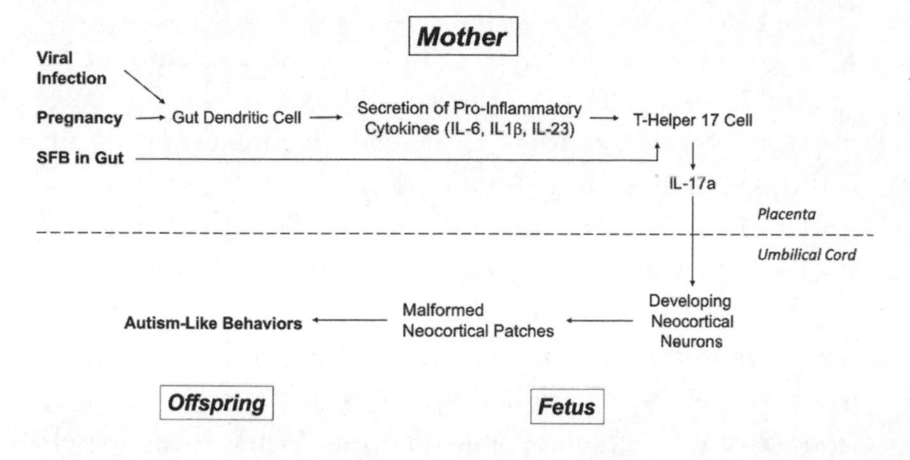

FIGURE 4. A molecular model for the contribution of maternal viral infection to fetal autism risk based on the work of Drs. Choi and Huh and their coworkers, as well as some other labs.

too early or too late, then the IL-17a surge produced by the infection will have no effect.[22]

Of course, there are some caveats to sound. The cortical patches produced in Huh and Choi's fetal mice are not exactly like the ones in humans with autism. And not all autopsy tissue from humans with autism reveals these cortical patches. And plenty of people have autism even though their mothers did not have a viral infection during pregnancy, so this IL-17a pathway is not the whole story for autism. Conversely, plenty of pregnant women get the flu and their children do not all have autism or schizophrenia. Nonetheless, these results make us think about individuality in a new way: both our mother's experience with infection during pregnancy and her complement of gut microbiota (and the way these factors interact over time) can potently influence our neuropsychiatric development.

═══

IT'S EASY TO IMAGINE that the effects of social experience on the development of individuality somehow occur in a different realm than those produced by the formative physical experiences we've discussed, like maternal viral infections. When we talk about early social experience we use words—like attachment, bonding, emotional warmth, and neglect—that are different from the biological terms, like IL-17a and dendritic cell. Let's be clear: these behavioral terms are important and useful, but they should not suggest to us that social experience operates in some special, woo-woo space where biology doesn't apply. When social experience—like parental neglect or bullying or nurturing—affects individuality in adulthood, it does so through biological effects on the brain. And when talk therapy works to ameliorate negative behavioral effects later on in life, it does so by changing the brain as well.

Here's an example. We know that children who don't receive regular loving touch in the first two years of life tend to have a broad array of lifelong neuropsychiatric problems, such as anxiety, depression, and intellectual disability. They also have a higher incidence of somatic (non-neuropsychiatric) illnesses, including persistent diseases of the gastrointestinal and immune systems. In recent years, a group of studies has shown that many different forms of early life social adversity—from a lack of parental loving touch to harsh, inconsistent discipline—produce exaggerated reactions to stress that persist through adult life and can contribute to these neuropsychiatric and somatic conditions. At least a portion of this enhanced stress reactivity comes from methylation in a region of DNA that suppresses the expression of the glucocorticoid receptor gene in certain brain regions.[23] Through a hormonal feedback loop, this increases the secretion of a key stress hormone, CRH, by neurons in the hypothalamus brain region, producing widespread biological effects. While methylation of the glucocorticoid receptor gene is only a part of the story of how social adversity in early life affects individual traits in adulthood, it is an important example. It shows that we can potentially understand the molding effects of early social experience in terms of identified molecular and cellular signals.

When Barbra Streisand's adored Coton de Tuléar dog, Samantha, was near death in 2017, the famous singer was devastated at the impending loss of her loyal and loving companion. So, as befits a wealthy star, she had her vet take tiny biopsy samples from Samantha's belly skin and cheek and sent them off, together with $50,000, to a company in Texas called ViaGen Pets. Using techniques originally developed at Seoul National University in South Korea, the ViaGen scientists were able to produce puppies from those cells that were genetic clones of Samantha. Streisand is now raising two of these puppies, which she has named Miss Violet and Miss Scarlett. While Miss Violet, Miss Scarlett, and Samantha are all genetically identical, they are not perfectly identical in terms of either looks or temperament. "They have different personalities," Streisand said. "I'm waiting for them to get older so I can see if they have her brown eyes and her seriousness."[24]

That Streisand's two cloned puppies are not precise copies of either her original dog or each other should be no surprise. After all, human identical twins share the same sequence of DNA, and yet, even when raised together, they have certain differences in appearance and even more noticeable differences in personality. These differences have been codified in guidelines for forensic science. While the total number of ridges in a person's fingerprints is about 90 percent heritable,[25] when the exact pattern of ridges and whorls is examined, it is revealed that identical twins do not have identical fingerprints. Furthermore, identical twins don't smell exactly the same. Well-trained sniffer dogs can reliably distinguish between the body odor of identical twins, even if they live in the same house and eat mostly the same foods.[26] This is a general finding. Genetically identical twins, raised in the same household, will still show differences in both physical and behavioral traits. Perhaps the best illustration of this is that people married to one member of an identical twin pair rarely find themselves romantically attracted to their spouse's twin. And this lack of spark is mutual: few identical twins are attracted to their co-twin's spouse.[27]

So why aren't identical twin people (or dogs) raised together more similar than they appear to be? Recall that twin studies suggest three general factors that influence traits: heritability, shared environment, and non-shared environment. In the case of identical twins raised together, differences in the first two factors are near zero. Does that mean that non-shared experiences account for all of the differences in traits? Not really. The truth is that "non-shared environment" is a garbage bag of a term that includes factors most of us would not regard as experience at all.

One of these important factors is the inherent randomness in the development of the body, particularly the nervous system. This is not "experience" or "environment" as we typically think of it—as something that impinges on an individual from the outside, like social experience or a viral infection. Rather, developmental randomness is intrinsic to the individual. During development, the human brain is estimated to produce about two hundred billion neurons, of which about one hundred billion survive competitive pruning during early life. In the adult brain, each of these surviving one hundred billion neurons makes about five thousand synaptic connections with other neurons. Those five hundred trillion synapses are not made randomly. The signals from the retina must be conveyed to the visual processing regions of the brain, and the signals from the parts of the brain that initiate movement must find their way to the appropriate muscles, and so on. The biological challenge is that the wiring diagram of the human brain is so enormous and complicated that it cannot be specified exactly in the sequence of an individual's DNA.[28] Subtle, random changes in the number, position, biochemical activity, or movement of cells within the developing nervous system can cascade through time to produce important differences in neural wiring and function between genetically identical twins raised together. Neurogeneticist Kevin Mitchell nicely sums this situation up by noting, "If you or I were cloned 100 times, the result would be 100 new individuals, each one of a kind."[29]

One might imagine that the differences between genetically identical people come about as a result of non-shared experience or developmental randomness affecting the timing or pattern of gene expression in various cells of the body. Indeed, this is sometimes true. For example, one of the chemical processes that regulates gene expression is the pattern of DNA methylation and the transfer of chemical acetyl groups (C_2H_3O) to histone proteins (histone acetylation). When we compare these processes in identical twin pairs, we find that the twins are very similar in early life, but that older identical twins accumulate more and more of these epigenetic differences as they age, causing their gene expression profiles to slowly drift apart.[30] That's really compelling, and so one might be tempted to imagine that regulation of gene expression is the whole story when it comes to understanding how experience drives individuality. But it's not. There are other important aspects of individual experience that are completely independent of regulated gene expression.

═

Like all biologists of my era, I was taught that somatic cells (all cells in the body except eggs and sperm) are genetically identical. In this way, the differences between cell types arise from varying patterns of gene expression, as determined by the process of development and by experience. That's what makes a liver cell different from a skin cell, even though they presumably both have the very same sequence of DNA.

Until recently, reading someone's DNA required a goodly amount of it: you'd take a blood draw or a cheek swab and pool the DNA from many cells before loading it into the sequencing machine. However, in recent years it has become possible to read the complete sequence of DNA, all three billion or so nucleotides, from individual cells, such as a single skin cell or neuron. With this technique in hand, Christopher Walsh and his coworkers at Boston Children's Hospital and Harvard Medical School isolated thirty-six individual neurons from three healthy postmortem human brains and then determined the complete genetic sequence for each of them.[31] This revealed that no two neurons had exactly the same DNA sequence. In fact, each neuron harbored, on average, about 1,500 single-nucleotide mutations. That's 1,500 nucleotides out of a total of three billion in the entire genome—a very low rate, but those mutations can have important consequences. For example, one was in a gene that instructs the production of an ion channel protein that's crucial for electrical signaling in neurons. If this mutation were present in a group of neurons, instead of just one, it could cause epilepsy. Another was in a gene linked to higher incidence of schizophrenia. There's nothing special about the brain in this regard. Every cell in your body has accumulated mutations, and therefore every cell has a slightly different genome. This phenomenon is called mosaicism, and when it occurs in cells other than sperm and eggs it is called somatic mosaicism. Sometimes somatic mosaicism is obvious. For example, the famous port-wine mark that adorned the head of Soviet leader Mikhail Gorbachev resulted from a spontaneous somatic mutation in a single progenitor cell, which then divided to give rise to a patch of cells forming enlarged blood vessels, thereby darkening that bit of skin.

Life begins as a single cell: a newly fertilized egg with a single genome. During development, both in utero and in early life, cells divide and divide again (figure 5). Early cells are multipotent: a single cell in a sixteen-cell embryo will have its progeny contribute to many different tissues of the body. As time goes on, cells and their offspring can become more restricted in their fate, giving rise to just skin cells or just brain cells, for example. In the end, the body is composed of about thirty-seven trillion cells, all originating from that single fertilized egg. Some cell types, like skin cells, keep dividing throughout your life to replace ones that die. Others, like most neurons, reach a point in early postnatal life where they stop dividing.[32] Most somatic mutations that change single nucleotides occur when a cell is not dividing.[33] The mutations that happen during cell division tend to be more drastic, involving loss, duplication, or inversion of big chunks of chromosomes or even entire chromosomes.[34]

When Walsh and his collaborators looked at neurons from the same person, they sometimes found the very same mutation in several neurons. In some cases, that group of neurons was clustered together in the same part of the brain. In other cases, they were spread widely across brain regions. Mutations found in brain cells were also found in single cells from the heart, liver, and pancreas. It is likely that these situations result from early mutations. Later mutations were shared by a few neighboring cells and were less likely to change bodily functions (unless the mutations activated the cell-division pathway that causes cancer), while earlier

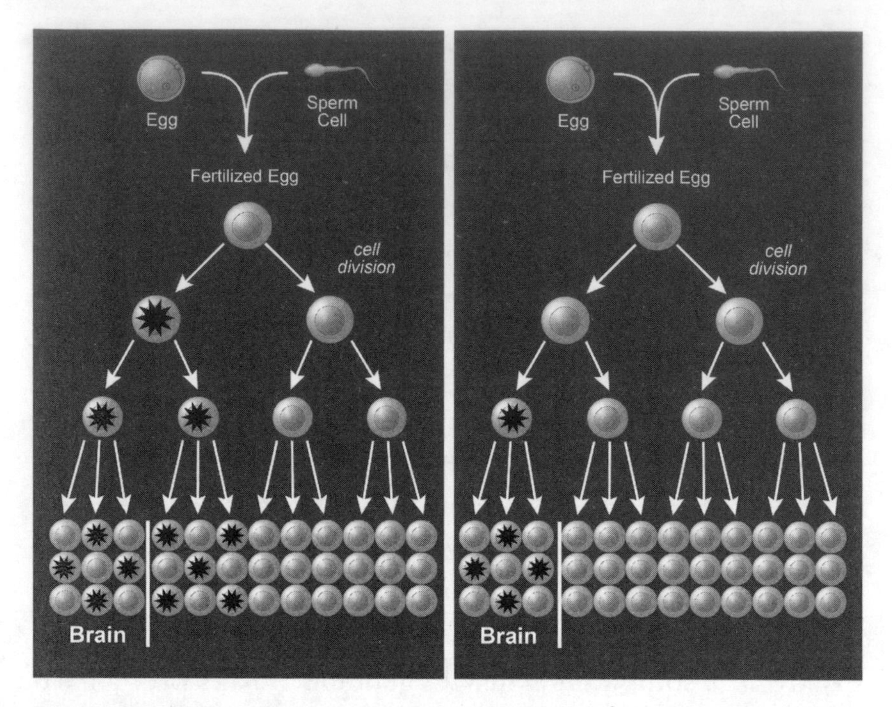

FIGURE 5. A spontaneous random mutation occurring in one cell can be passed to all of its progeny through cell division, resulting in a somatic mosaic. The left panel shows a spontaneous mutation (asterisk) occurring early in development that is passed to cells in various tissues. The right panel shows a mutation occurring somewhat later in development, which is passed to fewer cells and restricted to one organ—in this case, the brain. © 2019 Joan M. K. Tycko. Adapted with permission from Poduri, A. et al. (2013). Somatic mutation, genomic variation, and neurological disease. *Science 341*, 1237758.

mutations were shared by many more cells, distributed more widely in the body.

At present, we don't have a very big data set of fully sequenced individual human neurons, and the ones we have are mostly from autopsy tissue. We know of a few cases where spontaneous somatic mutations have given rise to severe neurological diseases—such as an overgrowth of one cerebral hemisphere, called hemimegalencephaly.[35] It's almost certain that a fraction of heretofore mysterious neurological diseases, like epilepsy of unknown origin, results from spontaneous somatic mutations affecting the electrical function of a group of neurons. It's also likely that somatic mosaicism contributes to individual differences in cognition or personality that do not rise to the level of disease. Stated another way, a portion of your individuality results from each cell in your body rolling the dice over and over again as you develop, grow, and age. These random changes, because they are in your somatic cells, not your eggs or sperm, will be unique to you and not passed down to your children. This distinction underlies an important point of terminology. The terms "genetic" and "hereditary" are often used interchangeably, but this is incorrect. Somatic mutations are genetic changes but—because they are neither inherited nor passed down to offspring—they are not hereditary.

So, we're actually a collection of thirty-seven trillion cells, each with a somewhat different genome. That's pretty hard to imagine. It's not just that it takes a village to raise a child. Each child *is* a village—or rather, a huge metropolis—of related but genetically unique individual cells. But it gets even more complicated. Sometimes the metropolis admits immigrants.

IN 1953, LONG BEFORE analysis of DNA was possible, a curious report was published in the *British Medical Journal.*[36]

> Mrs. McK., a donor aged 25, gave her first pint of blood in March of this year. When the blood came to be grouped, it seemed to be a mixture of A and O cells, for anti-A serum caused large agglutinates to appear to the naked eye, but the microscope showed these agglutinates to be set in a background of unagglutinated cells. The appearance was such as might be seen for a time after a large transfusion of O blood into an A recipient: but Mrs. McK. had never been transfused.

Now here was a mystery. How could Mrs. McK. have two different blood types running through her veins? Careful replication of the result showed that it wasn't due to simple contamination in the lab. One possible explanation was that Mrs. McK.'s conception resulted from a very rare situation in which a single egg is fertilized by two different sperm cells and develops into a single individual. But such dispermic people always have some degree of asymmetry in their bodies, like one ear noticeably larger than the other or different colored eyes. Mrs. McK. was normally symmetrical. Then, one of the doctors thought to pose an important question. "When asked if she were a twin, Mrs. McK., somewhat surprised, answered that her twin brother had died of pneumonia, 25 years ago, at the age of 3 months."

The explanation for Mrs. McK.'s double blood type is that the placenta is not a perfect barrier to the flow of cells. Some of her brother's type-A cells had passed into her body in utero and had replicated and survived for twenty-five years. At that age, about one-third of her blood cells were derived from her twin. When her blood was analyzed in the years to follow, the fraction gradually decreased but was never entirely eliminated—an odd kind of immortality.[37]

When cells from two different individuals become mixed, it is called chimerism. Recent work has shown chimerism to be widespread.[38] In fact, we're all chimeras, because cells readily pass from mother to fetus. In some cases, the maternal cells are cleared during childhood, but in others, the cells take up residence in a wide variety of organs and can persist for decades. Moving in the other direction, every woman has fetal cells in her body during late pregnancy, and about 75 percent of women still have fetal cells distributed throughout their bodies many years later.[39] In one recent study of autopsy tissue, 63 percent of women who had given birth decades earlier (their median age was seventy-five) had fetal cells in their brains.[40] It's worthwhile to note that fetal-to-maternal cell transfer can happen even in the case of miscarriage or abortion. There are women who don't even realized that they miscarried early in pregnancy, yet are still chimeras from invasion by those early fetal cells.[41]

Cell transfer across the placenta is a potential source of individuality, but we know very little about how nonself cells function in the body. Those fetal cells in the maternal brain can become electrically active neurons embedded within larger circuits. But it remains unclear if they matter for mental function and behavior. We don't know whether those invading fetal cells change a woman's experience of the world. When I was growing up in the 1970s, the mother of a friend liked to drink her coffee out of a mug emblazoned with the phrase "Insanity Is Hereditary: You Get It from Your Kids." Maybe she was right, but in a different way than she imagined.

Fetal-to-maternal cell transfer can be both harmful and beneficial.[42] At least some of the fetal cells that invade the mother's body are stem cells—undifferentiated cells that can ultimately become any type of cell. In some cases, the mother's immune system attacks these cells, producing autoimmune diseases like systemic sclerosis, which can damage the mother's skin, heart, lungs, and kidneys. In other women, the fetal stem cells can sometimes produce miraculous repair.[43] In one case study, a mother with a failing thyroid saw spontaneously renewed thyroid function. When a biopsy was performed, the cells of the regenerated thyroid were male, presumably seeded by a stem cell transferred from her son in utero. A similar case was reported for a mother's spontaneously remitting liver disease, but in that case the regenerating liver cells were from a pregnancy that was terminated.

═══

We have discussed several ways in which experience, broadly considered, drives individual traits: First, experience-driven regulated gene expression, in which stimuli such as temperature, social interaction, and birth season—acting epigenetically through transcription factors, DNA methylation, and histone modifications—determine which genes are turned on or off at various times in various cells. Second, somatic mosaicism, where random mutations accrue in the (non-sperm, non-egg) cells of your body to change their individual DNA sequence. Third, chimerism, when cells from another person invade your body. Finally, we have discussed the randomness of the development of the body and the brain, from the moment of conception to adulthood, that produces individual variation but is not really experience, in the sense that it is not a process that impinges upon your body from the outside.

Importantly, these are all mechanisms by which genetically identical twins can have divergent traits. Even side by side in the womb, twins do not have identical histories of developmental randomness, somatic mutation, experience, or chimerism. And, of course, the experiences, development, and accumulated somatic mutations of identical twins will continue to diverge after birth.

In recent years, there's been a lot of media attention about a phenomenon that scientists call transgenerational epigenetic inheritance, and which the popular press has mostly called "you can inherit your grandmother's trauma." The idea is that if your grandmother (or grandfather) lived through some physically or emotionally traumatic experience, like the 1918 pandemic flu, her trauma could be passed down through epigenetic changes (such as DNA methylation or histone acetylation) to her offspring. These changes would cause her offspring to experience some consequences of trauma (for example, anxiety, overeating, or high blood pressure) and the epigenetic changes could then be passed through the same mechanism down to you. Just to be clear, the mode of transmission here is thought to be epigenetic (modification of the patterns of and timing of gene expression), not genetic (modification of the DNA sequence itself as happens in the mutation and selection of conventional evolutionary change). And the mode of transmission is not merely intergenerational, passing from parent to offspring and then ending, but rather transgenerational, passing through at least two generations.

As of this writing, there are over fifty scientific papers claiming transgenerational epigenetic inheritance in humans. Some of the most cited are a group of reports from the rural Swedish region of Överkalix, which, over the years, has suffered from poor harvest and intermittent famine. These reports found that Överkalix grandsons lived longer if their grandfathers lived through famine in their prepubescent years. But the granddaughters of women who had survived famine had lower life expectancy.[44] The authors write, "We conclude that sex-specific, male-line transgenerational responses exist in humans and hypothesize that these transmissions are mediated by the sex chromosomes, X and Y."

I won't burden you with the details, but I'm sad to say that there's not a single one of these fifty-plus epidemiological studies that I find convincing. They tend to suffer from inadequate sample size, poor statistics (which do not correct for multiple comparisons), and hypothesizing only after the results are known.[45] The few studies in humans that have sought to actually measure epigenetic marks across generations—for example, in human sperm cells—have suffered from many of the same methodological problems.

For transgenerational epigenetic inheritance to work, the epigenetic modifications of your grandmother's DNA that occur in her brain to produce anxiety must also be transmitted to her eggs so that they can be passed to the next generation. Then, these marks must somehow act on the brain and body to change expression in specific target cells to reproduce the same behavioral and somatic traits in the next generation. Then, of course, this whole process must happen a second time—from your mother or father to you.

There is no evidence to show that any of these steps occur in humans. The long-standing dogma in developmental biology has been that epigenetic marks on DNA and histone proteins are removed very early in development, at a point where any given cell in the developing embryo has the potential to become any type of cell in the body. Recently, it has been shown that there are a very small number of sites in the mouse genome where these epigenetic marks are not completely erased, and so these could potentially serve as a substrate for transgenerational epigenetic inheritance.[46] There are some other non-DNA inheritance mechanisms, involving RNA interference with gene expression, that operate in plants and worms, but they have yet to be demonstrated in mammals, much less in humans.[47] At present, I remain unconvinced by claims for human transgenerational epigenetic inheritance. As they say, "Extraordinary claims require extraordinary evidence," and such extraordinary evidence has not emerged. However, I'm not willing to slam the door on the possibility that such a mode of human inheritance might be convincingly and mechanistically demonstrated in some limited way in the future.

═══

THERE'S SOMETHING IN NATURE that loves individuality, even in a situation that's designed to squash it. This was demonstrated when Benjamin de Bivort and his colleagues at Harvard University took genetically identical fruit flies and raised them in the lab to have as similar experiences as possible. Then, they placed individual flies in tiny Y-shaped mazes and made videos as they explored.[48] Some flies had a noted preference for turning left and others for turning right. On average, there were about the same number of righties as lefties. This wasn't just a spur-of-the-moment thing. Righty flies preferred to turn right day after day, and the same was true for lefty flies. It wasn't an artifact of lingering odors, as mutant flies that couldn't smell also had consistently behaving righties and lefties. When the scientists bred righty flies together, their offspring had, on average, an equal preference for left and right turns. The same thing happened with the progeny of lefty fly pairs. These results indicate that the trait of turn preference is not heritable.[49]

Then the scientists looked at several different strains of fly, where each individual fly of a single strain was genetically identical but there were genetic differences among strains. All of the strains had nearly identical average turning bias: about 50 percent righty. But some strains of fly had a greater number of individuals with an extreme preference: they would nearly always turn right or nearly always turn left. This means that, while the bias of an individual fly to turn left or right is not heritable, the total amount of variability across the population is determined by genetics. One way to think about this is that there are genes in flies that don't determine right or left preference, but rather influence something like decisiveness (or maybe, at the risk of anthropomorphizing, stubbornness). The specification of a fly to be a righty or lefty is random, but once that has been set, whether that fly will display a mild or extreme preference is genetically influenced.

This is important because it implies that behavioral individuality is itself a trait that is subject to evolutionary forces. Genes that drive a wide range of individual behaviors can be nature's way of ensuring that there's enough variation that a population won't be entirely wiped out by a catastrophic event.[50] If, say, only extreme left turners or extreme shade seekers would survive some drastic perturbation of the environment, then complete annihilation of the group would be avoided if there were a few of these kooky eccentrics left to live and breed another day.

THREE

I Forgot to Remember to Forget You

EVERYONE HAS A UNIQUE LIFE STORY, AND EVERYone's story is wrong. Our memories of events are notoriously unreliable. Storing and retrieving autobiographical memory is not like writing in a book that is later opened to reveal perfect sentences. It's not even like taking a photograph and leaving it out in the sun to fade, with the fine details gradually obscured by time. Rather, our memories of events fail in particular, reproducible ways, even in those of us who pride ourselves on having good memories.

Put simply, memories are not objective *recordings* of events; they are the unreliable traces of our individual *experiences* of events. Two people, standing side by side, will have different experiences of the same event based on their prior life history. If I had a traumatic experience with fire in the past, then my experience—and hence my later memory—of seeing a house fire will be different than yours, even as we watch the fire engine roll up to the scene together. And the memory for an event can continue to change long after it occurs. After memories are stored in the brain, they can be altered, both by subsequent experience and the mere act of recollection. While memories formed by a particular life course are central to human individuality, their malleability makes it clear that our most strongly held beliefs about ourselves are continually and messily constructed and reconstructed.

On the morning of April 19, 1995, Timothy McVeigh parked a truck containing a huge bomb at a drop-off zone in front of the Alfred P. Murrah Federal Building in Oklahoma City. He lit a slow-burning fuse and walked to his escape car, parked a few blocks away. The fuse ignited a horrifically effective bomb, which had been constructed by McVeigh and his friend Terry Nichols. The bomb was made from stolen commercial blasting explosives, ammonium nitrate fertilizer, racing fuel, diesel fuel, and acetylene cylinders, all packed into barrels. The enormous blast collapsed the front of the building, and blew out window glass and damaged adjacent structures within a sixteen-block radius. One hundred and sixty-eight people were killed, including nineteen children, most at an on-site day-care center for federal employees that was situated directly above the bomb.

The FBI sprang into action and, within a few hours, one of the truck's axles, bearing a legible vehicle identification number, was found among the wreckage. This quickly led the feds to Elliott's Body Shop, a Ryder truck rental location in nearby Junction City, Kansas. They called the body shop to say that they were sending an agent over to interview the workers who had seen the rental transaction: owner Eldon Elliott, mechanic Tom Kessinger, and bookkeeper Vicki Beemer. All of them recalled that the man who had rented the truck two days earlier had given the name Robert Kling, but only Kessinger remembered another man with him. An FBI artist raced to the scene and, working with Kessinger, produced sketches of the two men (figure 6). Robert Kling was called John Doe #1 and his accomplice was John Doe #2.

FBI agents went door-to-door in Junction City, showing the sketches and searching for leads. They got a hit at the Dreamland Motel, where the owner identified John Doe #1 as a man who had checked in on April 15, stayed through April 18, and kept a Ryder truck parked outside his room. The owner recalled that, after briefly sputtering (presumably having forgotten his alias of Robert Kling), the man had given his name as Timothy McVeigh. When the agents typed McVeigh's name into police computers, they couldn't believe their luck. McVeigh was, at that very moment, held in jail in a small town about a two-hour drive north of Oklahoma City. His getaway car had no rear license plate, and when he was pulled over for this infraction, the local patrolman spotted a concealed pistol and arrested him. The address on McVeigh's forged driver's license was a farm in Michigan owned by Terry Nichols. A few hours later, the farm was raided and Nichols was arrested. Bomb-making supplies and a hand-drawn map of Oklahoma City were found, with the location of the Murrah Federal Building and McVeigh's getaway car marked in red ink.

This was fine investigative work, but there was one remaining problem. While John Doe #1 was clearly Timothy McVeigh, John Doe #2 didn't look anything like Terry Nichols. When Attorney General Janet Reno announced the arrests of McVeigh and Nichols on TV, she emphasized that

FIGURE 6. The eyewitness sketches of the presumed Oklahoma City bombers that were distributed by the FBI. John Doe #1 (Timothy McVeigh) is on the left and John Doe #2 is on the right. Used with permission of the FBI.

"John Doe #2 remains at large and he should be considered armed and dangerous."[1]

The eyewitness sketch of the Oklahoma City bombing suspects may be the most famous one in the history of American criminal investigation. It was in every newspaper and magazine and in constant rotation on TV news. John Doe #2 was said to have a tattoo of a snake on his left bicep and wear a baseball cap with blue and white markings. A $2 million price was put on his head. Leads poured in to the hotline established by the FBI, and over ten thousand agents and others were assigned to follow them up. By various accounts, John Doe #2 had been seen with Timothy McVeigh in an Oklahoma City strip club or running out of the Ryder truck just before it exploded or buying fertilizer, presumably to construct the bomb. None of these stories could be verified. Fourteen men who resembled the sketch of John Doe #2 were taken into custody, but all were released with solid alibis. After many weeks, it became clear that the most intensive manhunt in the history of the FBI had failed.

It is almost certain that the search failed because there was never anyone with Timothy McVeigh when he rented the Ryder truck. Later, it was discovered that the day after McVeigh's visit, two men came in to Elliott's Body Shop to rent a truck. They were US Army Sergeant Michael Hertig, who, like McVeigh, was blond and fair skinned, and his friend, Private Todd Bunting, who was dark haired, muscular, and a dead ringer for John Doe #2. Tom Kessinger, the mechanic, had, despite the best of intentions, committed an error of memory. He correctly described the features of the innocent Todd Bunting but attributed his presence to the McVeigh episode that had occurred on the previous day. This merging of separate incidents is one of the typical ways in which memory fails us in daily life.

IT HAPPENS EVERY HOUR in police stations around the world. A crime has been committed, and a suspect is paraded in front of an eyewitness in a lineup with several others, often of the same general appearance. In this situation, if the real perpetrator is not among those in the lineup, some eyewitnesses, mostly acting without malice, will pick the person who most closely matches their memory of the perpetrator. The accuracy of eyewitness identification is not improved when a "six-pack" of mug shots is used in place of a live lineup. These lineups have been the cause of many wrongful convictions over the years.

It's hard to know the true prevalence of convictions based on mistaken eyewitness testimony, because most mistakes remain unrevealed.[2] We can make an estimate, however, based on simulated lineups in the laboratory. Experimental subjects were shown a video of a crime, in which the face of the "criminal" is clearly visible, and then presented with a lineup of six suspects, none of whom appeared in the video. In this situation, about 40 percent of the subjects nonetheless picked someone out of the lineup. Usually, but not always, this was the person who was the closest physical match to the criminal. If the subjects were told that others have already identified a particular suspect in the lineup and they just need to confirm or deny it, the rate of false recollection rises to 70 percent. Furthermore, when the subjects who picked a suspect were asked how confident they were in their identification, most said that they were absolutely certain.[3]

═══

OUR AUTOBIOGRAPHICAL MEMORIES ARE subject to all kinds of distortions—what psychologist Daniel Schacter cheekily calls "sins of commission."[4] In addition to misattributions of time, as in the case of John Doe #2, and the suggestibility of eyewitnesses, there's bias, which is the warping of one's recollection to mold to present beliefs, knowledge, and feelings. For example, after a bad breakup, people's recollections of the early stages of the relationship, previously recalled with pleasure, often turn darker. Or people will say, "I always knew that candidate X would win the election," even when they had voiced doubts about that outcome beforehand.

Some of the ways autobiographical memory fails are well known. Generally, our memories of recent events are more accurate and detailed than our memories of the distant past. But there are other, less obvious changes. If I ask you to recall a recent event, you are most likely to imagine it from your own point of view, with the camera, as it were, in your own eyes. This is called field memory. But if I ask you to recall a memory from your childhood, there is a much greater probability that your point of view will shift to that of an observer; you will see yourself in the scene rather than seeing the event though your own eyes. Furthermore, if asked to recall the emotional tone of a past event, you are more likely to evoke a field memory, while if asked to recall facts of an event, you are more likely to call forth an observer memory. The key point here is that the way we recall the memory is not set in stone. It can be strongly influenced by the task at hand.[5]

Another time related phenomenon is that repetition of experience renders memories generic. If you've only been to the beach once, then you are likely to remember many details of that experience. But if you've been over fifty times, you're unlikely to remember details of visit number thirty-seven, unless something emotionally affecting occurred. Perhaps visit number thirty-seven was the day a dead whale washed up on the beach, or the day you met your future spouse. Then the details of that day would likely be written into your memory deeply and retained with greater detail and fidelity. Emotions, both positive and negative, are the currency of autobiographical memory. Emotions cause the brain to store memory in a stronger and more permanent fashion, set down in bold type and italics. This reinforcement of emotional memories is mostly good and sometimes bad. It's good because emotionally salient events are often the ones you most need to remember later in life. However, in some cases, memory can become pathologically persistent, as when the memory of a traumatic experience—like an assault or a soldier's time in combat—is recollected incessantly.

If our memories for events are often so inaccurate and changeable, then why do we even have them? What is memory for? The main answer is that memory allows us to learn: to adjust our behavior based on individual experience and therefore efficiently find food, avoid predators, find and attract mates, and so on. In other words, memory does for the individual what evolution of the genome does for the species over many generations: it allows us to respond to the environment in a way that increases the chance of surviving and passing genes on to the next generation. That is endlessly useful. For example, a newborn mouse has an inborn fear of foxes, even if it is the descendant of many generations of lab mice with no exposure to foxes at all. This is a useful adaptation for mice in the wild, but it is not a good general strategy for dealing with a changing world. It is not possible to encode all useful behavioral responses into the genome in order to have a newborn equipped to deal with every eventuality. It's both more efficient and flexible to have animals remember and learn, even if they don't do so perfectly. And there is another benefit. The act of recollection allows us to mentally time travel to a past event, and this allows us to imagine a future as well as a past. Memory releases our mental life from the tyranny of the present moment. And imagining a future allows us to make predictions, which is a requirement for decision-making.

Another answer to the question of what memory is for is that the particular failures of autobiographical memory are actually features rather than bugs. For memory to be useful, it must be updated and integrated with subsequent experience, even if it alters the memory of the original event. In that way, it's helpful for recollection to render the memory of an event malleable, so that it may be integrated with the present. In most situations, a generic memory compiled from many trips to the beach is more useful in guiding future decisions and behavior than fifty stand-alone, detailed, and accurate beach trip memories. The repetition-driven loss of detail allows for the efficient use of the brain's limited memory resources.

In other words, it's not surprising that our memories for events are often inaccurate, because the particular way in which memories are compromised is often useful. What's surprising is that we mostly fail to recognize this in our daily lives. We humans all have an inborn tendency to create a plausible story out of memory fragments. Because of this ongoing narrative construction, we are often confident about the veracity of blurred memories and allow them to form the basis of our core beliefs about ourselves.

IN ADDITION TO MEMORIES of events, we also store memories of facts or concepts that are not linked to any particular event. For example, I can tell you that the capital of Mongolia is Ulaanbaatar, even though I cannot recall where or when I learned that fact. Or, I might have the fact right but get the source and time wrong—recalling that I learned it in high school, forty years ago, when I really read it on Wikipedia just last year. Likewise, I might be able to explain the concept of transitivity in mathematics without remembering when or how I learned it.

Psychologists call this decoupling of facts or concepts from events "source amnesia," and everyone has it to some degree, although it becomes more prevalent with normal aging.[6] On average, our memories for both facts and concepts, while not perfect, are usually more accurate than our memories for events. This may reflect that they are less rich in detail and context than events, which involve all of the senses. Facts and concepts are, in a way, already distillations of raw experience.

Bringing these concepts down to the level of individuals, we may say that "Fred has a bad memory" or "Sally has a good memory." But, of course, memory is not a single phenomenon. Not only are some people good at remembering facts and concepts but poor at events, but there is considerable variation in people's ability to remember particular types of facts and concepts. We all know people who have an impressive recall for comedy routines but struggle to commit music to memory. Others might be poor at putting names to faces but great at remembering what they've read. And among those people with a good memory for written material, there are different strategies involved. Some may recall a visual image of the page with the text laid out, while others recall the sounds and meanings of the words without a corresponding image of the page.

Memories for events, facts, and concepts all fall into a category called explicit memory: specific information that can be brought to mind with conscious mental effort. Explicit memory is what we usually mean when we talk about memory in everyday conversation. However, there's another type of memory that is equally important but less discussed. That is implicit memory: memory that is acquired and used subconsciously, without mental effort. Implicit memories are mostly acquired through repeated practice rather than from a single event.[7] Generally speaking, implicit memories are more stable than explicit ones and are stored in different brain circuits. That is why you may have to hunt around the house to find where you left your wallet (an explicit memory), but you're unlikely to forget how to ride a bike (an implicit memory). Our individuality is formed as much by our subconscious implicit memory as it is by the explicit memory for facts, events, and concepts that gets so much attention.

Figure 7 summarizes many years of work attempting to categorize types of long-term memory. The evidence for making these distinctions often comes from the analysis of patients who have sustained brain damage. For example, people with injuries to the medial temporal lobe can have

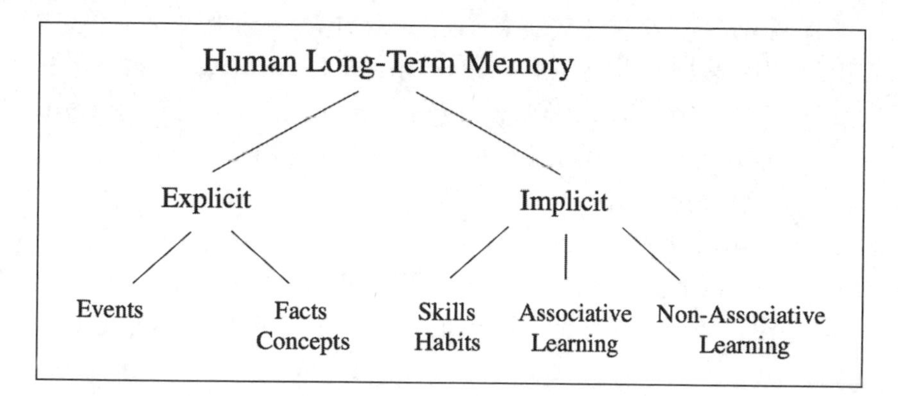

FIGURE 7. A taxonomy of human long-term memory. Explicit memory includes memories of events (autobiographical memory) or facts and concepts (semantic memory). Implicit memory is acquired and used without conscious attention, yet can guide decisions and behaviors. It includes memory for skills and habits (procedural memory), as well as simple associative learning (like eyelid conditioning) or nonassociative learning (like habituation of the orienting response).

a profound anterograde amnesia—an inability to form new memories, starting when the injury was sustained. They also display some degree of retrograde amnesia, which is the erasure of memory for a period of months or years before the injury, leaving older memories intact.[8] Originally, it was thought that medial temporal lobe amnesiacs had no ability to form new memories at all. But over the years, it has been revealed that the ability to form new implicit memories remains intact.

Reading text reflected in a mirror is a task that is difficult at first but can be gradually improved with practice. If someone with medial temporal lobe damage practices mirror reading for thirty minutes a day for three days and is then tested on mirror reading on the fourth day, their speed will have improved. But if you ask them if they have ever tried mirror reading before, they will say no. They will have no memory of the task, the room, or the person who helped them.[9] Recalling the *event* of training for mirror reading is a form of explicit memory that requires intact medial temporal lobe circuits. But improvement in mirror reading is a *skill*, and hence a form of implicit memory that can be stored and recalled even when these circuits are damaged. While mirror reading is a cognitive skill, motor skills, like improving one's tennis swing with practice, are also preserved in patients with temporal lobe amnesia.[10]

Another type of implicit memory is eyelid conditioning, a type of associative learning. If I play a soft tone, you will not blink your eyelid in response to it. But if I direct a brief puff of air to your cornea, you will blink reflexively. The blinking is not a conscious decision; it will happen whether you intend to blink or not. Then, if I pair the tone with the air puff—such that the start of the tone precedes the start of the air puff, and they end together—and I repeat this pairing many times, you will gradually learn that the tone will predict the air puff. The result is that you will begin to blink your eye earlier, so that your eyelid is at least partially closed and your cornea is protected when you expect the air puff to arrive. Again, eyelid conditioning is subconscious. You cannot override it or acquire it faster with force of will. It will happen no matter what you do.

Listen: Billy Pilgrim has come unstuck in time.

FIGURE 8. Mirror reading is a skill that can be improved with regular practice. Even people with brain damage who have profound anterograde amnesia for facts, events, and concepts can still improve in a mirror reading task. Here, the two-axis reversed text is a famous line from Kurt Vonnegut's 1969 novel *Slaughterhouse-Five.*

An even simpler form of subconscious learning is habituation, a type of nonassociative learning. If I were to stand just outside your field of view and drop a book on the floor, you would rapidly swivel your head to investigate. This behavior is called the orienting reflex, and it was first described by the Russian physiologist Ivan Sechenov in 1863.[11] The orienting reflex is a response to novelty, and so if I were to continue to drop the book, say, once per minute, you would soon learn to ignore it—to habituate—and the orienting reflex would be suppressed. Habituation is specific to a particular stimulus. A bright light will also elicit an orienting response. But repeated book drops will not diminish your response to the first bright light or vice versa.[12]

Interestingly, once you have habituated to some stimulus that's regularly repeated, then its absence becomes novel and can elicit an orienting reflex. To illustrate this phenomenon, the neurophysiologist Karl Pribram told the story of the Third Avenue elevated railway line, which once ran along the Bowery in New York City and was famously loud.[13] The train ran on a regular schedule through the night, and the residents of the apartments along the Bowery had become habituated to the intermittent noise. When train service on the line was stopped, in 1954, the police started receiving calls from local people who had been awakened out of a sound sleep by some occurrence they could not entirely define but which they assumed must be a prowler. An astute detective, who found no unusual evidence of creeping miscreants, soon realized that these calls were clustered at the times when the elevated train used to pass by. The strange occurrences that woke people were the orienting responses evoked by roaring silence when noise was expected. Once people habituated to the absence of elevated train noise, their panicked calls to the police subsided.

We like to imagine that we are fundamentally creatures of free will. We reliably call certain facts, events, and concepts to mind. We make conscious decisions and act volitionally. Our individuality is inextricably bound up with a deep sense of agency and autonomy. To a large degree, this is a trick our brain plays on us. Most of our behavior is subconscious and automatic. In the words of neuroscientist Adrian Haith: "Almost everything you do is a habit."[14] A habit is not just a behavioral routine that is formed and then performed at a subconscious level, it must become divorced from an ultimate goal. Your goal might be to stop off at the Thai restaurant for takeout after work, but instead you habitually drive straight home. The same habit can be either beneficial or detrimental, depending on the context. You may type rapidly and automatically on a standard QWERTY keyboard, but this habit will fail you if only a Dvorak keyboard, with a different layout of letters, is available.

Generally speaking, when you are learning a new task, your behavior is flexible and goal directed at the outset, but becomes automatic and habitual with repeated practice. For example, when you first learn to drive a car, you must think carefully about every action: steering, braking, signaling, scanning the road ahead. But with time, these actions become mostly automatic. Driving has become a habit and no longer requires your full mental attention.

While habits have the limitation of being inflexible, they have the advantage of being easy. The sad truth is that much of life is predictable and boring, so the inflexibility of habits is rarely a problem. Crucially, when behaviors become habitual, the conscious mind is free to ponder, predict, and plan. All of us have a collection of learned behaviors acquired over a lifetime of practice. We master one task over time, render it habitual, and then move on to the next one. In this way, we each assemble a vast library of habits and skills that can be called up automatically. As Haith writes, "Atop this massive conglomeration of habits sits a thin sliver of cognitive deliberation that steers only the highest-level decisions that we need to make." Without habits, our brain would be instantly overwhelmed with a multitude of tiny decisions better left to rapid, automatic processes.

These examples of subconscious learning that we've explored—including the learned skill of mirror reading, habituation of the orienting reflex, and associative eyelid conditioning—all rely on implicit memory and so are spared in patients with medial temporal lobe amnesia.[15] Indeed, implicit memory mostly depends on circuits involving other brain regions.[16]

ALL MEMORIES, EITHER EXPLICIT or implicit, must be stored in the brain. The very short-term memory that you need to keep a phone number in mind as you dial is encoded by reverberating electrical activity that passes back and forth between three brain regions: the thalamus, the frontal cortex, and the cerebellum.[17] This working memory is also what you need to keep the beginning of a long sentence in mind as you read to the end. It is easily disrupted by competing mental activity (like someone speaking to you while you dial or read) and is discarded almost immediately after it is used.

Longer-term memories require more enduring alterations. Patterns of electrical activity associated with particular experiences must give rise to changes in the interconnected networks of neurons that make up the brain. Signaling in the brain has a mixed electrical and chemical character. Neurons convey information though rapid, all-or-none electrical signals called spikes. A spike travels down the long, thin, information-sending fiber of a neuron called the axon.

When the spike invades specialized active zones in the axon, it triggers the release of chemical neurotransmitter molecules. These diffuse across a tiny saltwater-filled gap and activate receptors on the information-receiving part of the next neuron in the signaling chain, called the dendrite, sometimes producing an electrical response in the next neuron in the network. These locations where neurotransmitters are released by one neuron and then received by another are called synapses.[18]

Let's play God for a moment. If you are the Great Engineer and you want to build memory storage in the brain, there are two main options. First, you could have experience-driven patterns of electrical activity persistently change the strength of chemical transmission across synapses. This could take the form of making synapses stronger (or growing new ones) or making synapses weaker (or eliminating existing ones). Together, these changes are called synaptic plasticity. Or, you could have experience alter the electrical signaling properties of whole neurons. For example, you could alter neurons to make making them more or less likely to fire spikes, or to fire spikes in different temporal patterns. These experience-driven processes are called intrinsic plasticity. It turns out that both intrinsic and synaptic plasticity are involved in storing long-term memories, although much more attention has been paid to the latter. Because each neuron in the brain receives an average of five thousand synapses, the information storage capacity of synaptic plasticity in much greater than that of intrinsic plasticity. Intrinsic and synaptic plasticity interact in complex and useful ways to store memories.[19]

Just as important is what's not changed to store memory. Experience does not modify the sequence of DNA in the cells of the brain, so this process cannot be the substrate of memory. Rather, memory is yet another example, albeit a specialized one, of how experience changes gene expression to produce lasting changes.[20] It's not unlike the example we discussed in chapter 2 about how the ambient temperature in the first year of life determines the degree of sweat-gland innervation. Only in this case, the tissue being altered by experience is not the peripheral nerves or the skin but the brain, and the changes in gene expression give rise to synaptic and intrinsic plasticity, the stuff of memory.

Much of the biology underlying the recollection of memory for facts and events remains poorly understood, but there are some general features we understand. Recollection of memory typically involves electrical activity in at least some of the neurons and synapses that were active during the original experience. However, the story is more complicated than that, as the neural circuits and brain regions involved in storing memories can sometimes shift over time. As mentioned earlier, people who sustain damage to a brain region called the medial temporal lobe will typically lose their memories for facts and events from a period of months or years prior to their injury, also known as retrograde amnesia. Older memories for facts and events remain intact, however, suggesting that they have been transferred from the medial temporal lobe to other brain regions.

In emotional situations, certain neurotransmitters (like dopamine and norepinephrine) and hormones (like epinephrine and corticosterone) are released. Some are released and act within the brain, while others are released in the body and make their way to the brain. These emotion-associated chemical signals can increase the extent of synaptic and intrinsic plasticity driven by experience, thereby strengthening memory. Importantly, this strengthening doesn't just happen at the initial time the memory is laid down. If the act of recalling a memory evokes emotional responses, then this chemical process can further strengthen (and warp) the memory every time it comes to mind.

IS MEMORY STORAGE IN the brain an unlimited resource, or can we run out of space? Can training in one particular skill or task crowd out our ability to excel at another, or is there room for limitless self-improvement? Disappointingly, there are some reasons to believe that memory resources are limited.

To become a licensed taxi driver in London, one must accomplish a prodigious feat of learning. The exam requires comprehensive memorization of the city's twenty-five thousand streets, as well as hotels, restaurants, landmarks, and the optimal routes between them, a body of information called "the Knowledge." Even after several years of study, many drivers do not pass the exam and must try again or give up. There was great excitement in the field when a careful study by Eleanor Maguire and her coworkers revealed that, on average, compared to age- and education-matched controls, licensed London taxi drivers have an enlarged brain region called the posterior hippocampus.[21] This region is thought to have a special role in the processing of spatial information. This result could mean that the intensive training performed in order to pass the exam caused the volume of the posterior hippocampus to increase as it came to contain a detailed mental map of the city. Alternatively, it could mean that those blessed with a large posterior hippocampus prior to training were more skilled at spatial cognition and hence more likely to successfully acquire the Knowledge and pass the exam.

A more recent study followed prospective London taxi drivers with repeated brain scans before and after their extensive bout of memorization. It showed that drivers who studied and passed the test significantly enlarged their posterior hippocampus, while those who failed or dropped out did not, and neither did the control subjects of similar age.[22] Thus, acquisition of the Knowledge appears to cause enlargement of the posterior hippocampus.

As the posterior hippocampus grows with spatial learning, this expansion comes at the expense of an adjacent brain region, the anterior hippocampus, a structure that is not involved in spatial cognition but rather in formation of new visual, but nonspatial, memories. This is likely to explain why, on average, London taxi drivers perform somewhat worse on visual memory tests than matched control subjects or drivers who failed to pass the exam. This finding indicates that at least some mnemonic and cognitive resources are limited in the brain and can be dynamically assigned to the tasks at hand by extensive training. Interestingly, retired London taxi drivers slowly return to the control state, with smaller posterior hippocampi, larger anterior hippocampi, improved visual memory scores, and a fading recollection of London's twisting streets.[23]

There are two reasons why London taxi drivers are considered an unusually good population for examining training-evoked changes in the brain. First, acquiring the Knowledge is a difficult task, but one that doesn't require unusually high intelligence. The average intelligence for London taxi drivers is about the same as that of the general population in the United Kingdom. Second, unlike musical or sports training, which often begins in childhood and hence conflates brain development with learning, learning the Knowledge only begins in adulthood, after the taxi drivers' brains are mature.

The London taxi driver result prompts the question: Is the enlargement of a brain region (and the related shrinkage of a neighboring brain region) a general property of intensive training during adulthood, or is there something special about the taxi exam? After two years of study, medical students in Germany must take a grueling comprehensive exam, the *Physikum*, which tests knowledge in chemistry, physics, anatomy, and biology. Students are allotted three months of daily study sessions to prepare. Arne May and his coworkers scanned the brains of medical students and matched control subjects before the three-month-long period of studying, a day after the exam, and then again three months after the exam. Over the learning period, the volume of three brain regions increased in the medical students compared to the control subjects: the posterior parietal cortex, the lateral parietal cortex, and our old friend the posterior hippocampus. The increases in these three brain regions were sustained when measured three months later. As with the London taxi drivers, there were also reductions in an adjacent structure, the occipital parietal lobe,[24] suggesting that competition for brain space following intensive training in adulthood is a general principle. Unfortunately, we don't know whether there were particular cognitive impairments that correlated with these shrunken brain regions in the medical students.

It's likely that the growth and shrinkage of brain regions produced by studying for exams is not as long-lasting as that produced by acquiring and then using the Knowledge. This was suggested by a study by Eleanor Maguire and her team that focused not on medical students but on working doctors, a group that must acquire and deploy a great deal of knowledge during years of extensive training. Unlike taxi drivers, doctors score better than average on intelligence tests. As a result, they were compared with a population that had the same intelligence test scores, but didn't have university attendance or other intensive training (like trade school). In this comparison, neither the posterior hippocampus nor any other brain region was found to be enlarged in doctors compared to the control group.[25] This result suggests that acquiring a lot of information over years of training is not sufficient to produce lasting changes in the gross structure of brain regions.

One clever way to search for rapid, training-induced brain changes in adults is to perform scans of their brains before and after teaching them to juggle. Researchers divided an age- and sex-matched pool of volunteers into two groups, jugglers and non-jugglers, and gave them all brain scans before training. The group of jugglers were given three months to learn a classic three-ball cascade juggling routine and sustain it for one minute. When the juggling group was tested soon after having mastered that skill, they showed expansion of two areas in the brain—the mid-temporal cortex (on both sides) and the posterior intraparietal sulcus (on the left side only)—when compared to the control non-juggling group. The expansion of these particular regions makes sense, because the former is involved in tracking the speed and direction of moving objects while the latter is involved in attention and sensory-motor coordination. After a break from practicing for several months, most people in the juggler group could no longer juggle on their first attempt and the enlargement of the brain areas was partially reversed.[26] This reversal is probably similar to what's happening with the doctors. They study hard briefly to pass their exams, but most would be hard-pressed to pass again in mid-career without brushing up. The expansion and shrinkage of brain regions with memorization—whether an implicit sensory-motor memory like juggling, or the explicit memory of the London street map or the *Physikum* exam—seems to last only as long as that knowledge is actively used.

In order for a brain region to grow larger, there must be a significant addition of cellular material. Given what we know about synaptic plasticity, a large part of this memory-associated expansion of brain regions must involve enlargement of existing synapses, as well as growth of new synapses and the dendrites and axons on which they are formed. It is also possible that entirely new cells are added to brain regions. Glial cells are continually produced by cell division in the brain and so could contribute to regional growth.[27] In some very limited parts of the brain, like the hippocampal dentate gyrus, which is part of a brain circuit for memory of facts and events, new neurons are formed after birth. But while it is clear that new neurons are created in the brains of adult birds and rodents, there is a robust debate about whether the adult human brain can form new neurons, or if that is limited to early life.[28]

It's important to highlight that seeing a brain region expand with memory storage is an extreme case produced only by sustained training. In most situations, memories are stored without overt changes in the size of brain regions. You can imagine that if experience adds or strengthens some synapses while removing or weakening others in a brain region, there will be no overall change in the volume of that region, even as memories are stored by changes in the function of its neural circuitry. This type of change can lead to functional plasticity, even in the absence of detectable changes in the size of brain regions.

A good example of functional plasticity comes from the study of serious musicians. When players of string instruments, like the cello or guitar, are placed in a brain scanner and compared to controls, it is revealed that more of their brain is devoted to touch sensation from the left hand but the brain space dedicated to the right hand is unaltered. Importantly, the structure of bulges and grooves on the brain is identical between musicians and controls. There's no overall growth or shrinkage of brain regions. Rather, one particular area, called the primary somatosensory cortex, has more of its volume dedicated to the left hand at the expense of processing touch sensations from other parts of the body.[29] This occurs because the left hand performs the highly dexterous fingering motions, which require finer touch sensation and motor control than the strumming or bowing performed by the right hand. Presumably there are some deficits in touch sensation on the left side of the body of string players, produced by expansion of the hand area and the consequent shrinkage of brain space dedicated to the rest of the body surface, but these have yet to be investigated.

═══

ALMOST ALL OF OUR instincts about memory are wrong. We feel like creatures of free will, with detailed and unlimited recall of those events that have helped to form us as individuals. In reality, most of our behavior is composed of learned, subconscious habits and skills with only a thin veneer of decision-making at the surface. Our recollection of specific events is unreliable and subject to further distortion every time we recall them. Our memories of facts and concepts are only marginally better. When asked about how confident we are about the veracity of a particular memory, our estimation bears no relationship to the truth. We feel like we can learn more and more with no limit, yet intensive training in one type of memory seems to degrade our ability to store some other forms of memory.

Our memories are suboptimal, and yet we hold them close. They feel true and important. They are central to our sense of individuality and agency. The mismatch between how we revere our memories and how often they fail is striking. Why we should feel more agency than we really have is an interesting and open question. I tend to see it as more of a feature than a bug. When we feel that we are in charge of our behavior, that we are making decisions based on accurate recollection, this allows for more rapid decision-making in those cases of the "thin sliver of cognitive deliberation," where it is truly required. In other words, when we don't have to stop and second-guess whether we are in charge, we can be decisive when it really matters.

FOUR

Sexual Self

WHEN WE LEARN THAT A BABY IS BORN, OUR FIRST question is almost always "girl or boy?" In every society, sex is a fundamental trait used to categorize individuals. It's the first box to check on every form. When we meet someone new, we cannot help but seek to ascertain their sex. It's a deep, subconscious drive that cannot be turned off. Sex is also the individual trait that we're least likely to forget. You may not recall whether a person named Terry who you met briefly at a party two years ago had brown or black hair or worked as an accountant or a marketer, but you are very unlikely to forget his or her sex. No matter your culture, religion (or lack thereof), or politics, sex matters to all of us. That's why people can get so upset when talking about it. Complicated and changing ideas about sex, including gender identities beyond the traditional female/male binary, can challenge the essence of who we are.

ON SEPTEMBER 22, 1938, an urgent message was transmitted from the police station in Magdeburg, Germany, to Berlin: "Women's European high-jumping champion Ratjen, first name Dora, is not a woman, but a man. Please notify the Reich Sports Ministry at once. Awaiting orders by radio." The Reich sport minister, Hans von Tschammer und Osten, didn't want to believe this news—a deep embarrassment for the German state—and called for his own doctors to examine Dora. But their finding was the same. Dora Ratjen, nineteen, who had competed for Germany in Hitler's showcase 1936 Berlin Olympics, and who had set a world record for the women's high jump at the European Athletics Championship in Vienna just a few days earlier, was, in truth, a man—at least by the standards of Germany in 1938. Under Tschammer und Osten's direction, Dora's gold medal was quietly returned, her world record was struck from the books, and she was banned from sports competitions for life.

MALES TYPICALLY INHERIT ONE X and one Y sex chromosome, while females usually have two X sex chromosomes. One crucial gene on the Y chromosome, *SRY*, codes for an important protein that, by affecting the activation of other genes, guides male-typical development beginning early in embryonic life. In the presence of the *SRY* gene product, two small blobs of tissue are instructed to become testes, which secrete the hormone testosterone. Testosterone (or its metabolite dihydrotestosterone) then binds to specific receptor proteins in cells and has widespread effects throughout the body. It is a key signal driving male-typical development of everything from the genitalia (during embryonic development) to the Adam's apple (much later, during puberty). In females, in the absence of *SRY*, other genes actively drive these same blobs of embryonic tissue to become ovaries, which secrete the key hormones estrogen and progesterone.[1] Importantly, although testosterone is present in two different surges starting in early fetal life, the secretion of estrogen is suspended from the period shortly after birth to puberty. This means that, during certain critical stages of development, the main hormone difference is higher levels of circulating testosterone (and some testosterone-like hormones, collectively called androgens) in most males and lower levels in most females. However, females are not completely bereft of androgens as, starting around age eight, the adrenal glands secrete low levels of testosterone (and dihydrotestosterone and androstenedione), and these androgens are also important for normal development in females. In males, estrogen has a role in normal development and adult function as well, although some of the details have yet to be worked out.

In most people, sex determination is straightforward. You inherit an X chromosome from your mother and either an X or a Y chromosome from your father. If a Y-carrying sperm cell fertilizes the egg, you'll be male; if it's an X, you'll be female. If you carry XX sex chromosomes, then you'll develop ovaries, a vagina, and a typical vulvar region as you grow in utero. Later, typical female secondary sex characteristics like menstruation, rounded hips, and breast growth will emerge at puberty. If you carry XY sex chromosomes, then you'll develop testes and penis in utero. Typical male secondary sex characteristics like a lowered voice, increased muscle mass, and body hair will emerge at puberty. However, there are several ways in which this process can become complicated, resulting in a range of conditions that we call intersex, defined as people born with sex characteristics, including internal and external genitals, that do not fit typical binary notions of male or female bodies. Most intersex traits are identified at birth, while others may not be discovered until puberty or even later in life.[2]

Rarely, intersex conditions result from chromosomal abnormalities. For example, Klinefelter syndrome—a condition in which all or some of the cells in the body have an extra X chromosome, yielding the pattern XXY—can sometimes produce an intersex condition. Individuals with XXY chromosomes have a penis and testes, but, in some severe cases, these organs are small and incompletely formed. XXY boys also have attenuated puberty, with reduced body hair and muscle mass and, sometimes, breast enlargement.

Most intersex conditions arise from alterations in hormone signaling in chromosomally typical (XX or XY) individuals. If an XY person carries a mutation that interferes with the function of the androgen receptor or its downstream biochemical signals, this results in a condition called androgen insensitivity syndrome. Depending on the severity of the mutation and its distribution in various cells of the body, the genitalia in XY individuals with androgen insensitivity syndrome can range from fully masculinized (rare) to fully feminized (most common). Secondary male sex characteristics like voice pitch, muscle mass, and body hair distribution can show a similar range of variation. A highly feminized person with XY chromosomes may have internal testes that secrete testosterone but may appear externally typically female, with a vulva and vagina. These individuals are almost always raised as girls with no suspicion of any underlying problem. At puberty, estrogen that is metabolized from testosterone secreted by the internal testes causes female-typical breast and hip development. Menstruation never begins because of the lack of ovaries or uterus, and often this is when the disorder is investigated and diagnosed.[3]

XY individuals carrying mutations in the gene encoding the testosterone-metabolizing enzyme called 5-alpha reductase have a particularly challenging intersex situation. This enzyme converts testosterone into the more active metabolite dihydrotestosterone, which is crucial for masculinization of the external genitals during fetal development. Its impairment results in external genitalia at birth that are either fully female-typical or of intermediate form. Many but not all people with 5-alpha reductase deficiency are raised as girls. Later, during puberty, there is a marked masculinization, and the affected individuals develop typical male features, including increased body muscle mass, deepening voice, testicular descent, and absence of breast development. In some cases, the external genitals resolve into a semi-functional penis.[4] Most but not all people with 5-alpha reductase deficiency who were raised as girls come to identify as male after puberty.

Another type of intersex condition, congenital adrenal hyperplasia, occurs in XX people in whom the adrenal glands secrete an unusually large amount of testosterone due to a recessive gene mutation.[5] Again, there is a range of effects, depending on just how much testosterone is secreted and a few other factors. In severe cases, both the external and internal genitalia become ambiguous, often with an enlarged clitoris and a shallow vagina. In milder cases, the genitalia are mostly female-typical, but male-typical secondary sex characteristics often emerge, including body hair, increased muscle mass, and suppressed menstruation.[6] To complicate matters further, there are some women who produce unusually large amounts of testosterone but also carry androgen-receptor mutations, so their extra testosterone has no biological effects. The overall conclusion: Traditionally, most cultural ideas have reinforced the notion that biological sex is a clear and immutable binary trait. But in about one in three thousand live births, nature does not draw such a bright line between male and female bodies.[7]

WHEN MRS. RATJEN GAVE birth to her fourth child on November 20, 1918, there was some confusion in the room. Later, recalling the birth, her husband, Heinrich Ratjen, said, "I was not standing at my wife's bedside during delivery. Rather I was in the kitchen at the time. When the child was born the midwife called over to me, 'Heini, it's a boy!' But five minutes later she said to me, 'It is a girl, after all.'" The child had ambiguous genitalia, a penis with a fissure and opening on the underside, and the parents didn't know what to do. So they followed the midwife's advice, named the child Dora, and raised her as a girl. Dora attended a girl's school and wore girl's clothing, but around age ten she began to question why she felt and looked like a boy. As she told the doctors who interviewed her later in life, her concern mounted as she failed to develop breasts or other female-typical secondary sexual characteristics during puberty. Dora's first ejaculation horrified her. She felt trapped by her situation and, because of the stifling social norms of her time and place, unable to ask questions or confide in anyone about her condition. Chromosomal and androgen-receptor testing was not yet developed, so we don't know the genetic details of Dora's condition. What we do know is that, despite being raised as a female, Dora felt like a male and had a body that was mostly, but not entirely, male-typical.

Fearing detection, Dora avoided dancing or swimming but soon found some solace in her love of sport. By age fifteen she was a regional high jump champion and a contender for the 1936 German Olympic team. When the Nazi government passed over the top German women's high jumper, Gretel

Bergmann, who was inconveniently Jewish, this created a spot on the team for Dora. She may have appeared unusually deep voiced and lean, but her fellow athletes never suspected her secret. Years later, Gretel Bergmann recalled, "In the communal shower we wondered why she never showed herself naked. It was grotesque that someone could still be that shy at the age of 17. We just thought—She's strange. She's odd."

Dora finished fourth at the Berlin Olympics, just out of the medals, but continued to improve in the years to follow, breaking the women's world record in high jump just two years later. It was on a train home from that victory in Vienna that her secret was revealed. A conductor on the train suspected that Dora was a man dressed in women's clothing (cross-dressing was illegal in Germany at that time). When the train stopped in Magdcburg, thc conductor alerted a policeman to his suspicions and the policeman confronted Dora. After a brief denial, and a presentation of her women's ID from the recent European championships, Dora admitted that she had always felt like a man, a fact that was then confirmed by medical examinations. "Ratjen openly admits to being happy that the cat has been let out of the bag," stated a police officer after her arrest. Although fraud charges were initially filed against Dora, they were later dropped when the prosecutor concluded that there was never any intent to deceive, just a terrible misunderstanding caused by well-meaning but confused adults when Dora was a newborn. Dora's name was changed to Heinrich and he lived out the rest of his life, quietly, as a man.

The case of Dora Ratjen is similar to those in the 1960s and '70s, in which boys born with penile malformation received gender reassignment surgery as infants and have been raised as girls from the time of birth. The misguided theory that drove this ill-fated decision held that infants are a blank

FIGURE 9. Dora Ratjen at a high jump competition in 1937. Photo from the Bundesarchiv, Bild 183-C10379. Used with permission.

slate, and so chromosomal males could be raised to feel female. In fact, this completely failed. As they grew up, almost all reassigned boys reported that they felt male and almost all grew to become sexually attracted to women.[8] As a result, the medical community has changed the standard of care from encouraging parents to choose a gender for intersex children soon after birth to encouraging them to wait until the children express a clear gender identity.

DORA RATJEN'S CASE WAS one of several cited by the International Olympic Committee (IOC) when they began mandatory screening of athletes seeking to compete in women's events. The rationale for this practice has always been to catch male athletes masquerading as women. Remarkably, this has never happened.[9] Instead, screening has entirely served to humiliate and exclude people with intersex conditions.

The athletes called the first mandatory femininity screening, which began at the 1966 European championships, "the nude parade." In this situation, women who did not appear to the assigned panel of male doctors to be entirely female-typical could be called out of line and be made to spread their legs for closer examination. There was never any screening for athletes seeking to compete as male. In 1968, in response to complaints by female athletes, this degrading practice was replaced by a cheek swab to collect cells for chromosomal testing. The new rule held that only persons carrying XX chromosomes could compete as females. Not surprisingly, because sex is determined by a confluence of chromosomal and nonchromosomal factors, this method had problems.

One famous case was a Spanish hurdler named Maria José Martínez-Patiño, who had XY chromosomes and a profound androgen insensitivity syndrome. Her face and body were externally female-typical. She had breasts, a vulva, and a vagina, but no uterus or ovaries. She had always felt herself to be female and was raised as female. Her androgen insensitivity mutation assured that her body could not be affected by the testosterone produced by her internal testes. When her chromosome test was publicized, the response was immediate and brutal. Her medals and records were revoked and she was thrown off the Spanish team, losing her living allowance and her apartment. Her boyfriend left her and strangers pointed at her in the street. Later, she wrote: "If I hadn't been an athlete, my femininity would never have been questioned. What happened to me was like being raped. It must be the same sense of violation and shame. Only in my case, the whole world watched."[10] She appealed her case, arguing correctly that her body received no competitive advantage from androgens produced by her internal testes. Eventually she won, but the process took three years, and by then her hurdling career was over.[11] The strict XX chromosomal standard to compete as a female had clearly failed.

That said, there are still questions about advantages conferred in females who are XY with complete androgen insensitivity syndrome. This is suggested by the observation that the incidence of this syndrome in XY individuals is about 1 in 20,000 in the general population but about 1 in 420 in elite female athletes competing at the Olympic Games.[12] We know that *SRY*-driven development of the testes and subsequent testosterone production is far from the only consequence of having a Y chromosome.[13] That chromosome has about two hundred genes, of which at least seventy-two have been confirmed to direct the production of proteins. Some of those genes may confer advantages over XX athletes in certain sports in a manner that is independent of testosterone, perhaps by increasing height or lean muscle mass.[14] But, at present, there is no evidence that female XY athletes with complete androgen insensitivity syndrome develop any physical attribute important for athletic performance that is not also present in at least some XX females.

In 2013, the International Olympic Committee announced a new rule: athletes seeking to compete as female could do so only if their blood testosterone levels were below ten nanomoles per liter, with exceptions for cases of androgen insensitivity. If an athlete wanted to compete and exceeded the ten nanomole limit, then she would either have to have surgery (to remove internal testes), take androgen-suppressing drugs to meet the cutoff, or compete as a man. Only 0.01 percent of women have natural testosterone levels exceeding ten nanomoles, so one might imagine that few athletes would be affected by this rule.[15] However, the fraction of elite female athletes exceeding the IOC's testosterone standard is about 1.4 percent, 140 times higher than the general population. This suggests that naturally high testosterone might indeed confer an advantage for some female athletes who do not also have androgen insensitivity. In recent years, several elite female athletes have been barred from competition on this basis, including Caster Semenya, a middle-distance runner from South Africa, and Dutee Chand, a sprinter from India. Chand appealed her ban to the Court of Arbitration for Sport in Lausanne, Switzerland. She argued that she was born and raised as a woman and that she had not doped or cheated in any way. Why should she be forced to undergo surgery or take drugs in order to compete as a woman?

Testifying in support of the IOC's testosterone standard, champion British marathon runner and sports official Paula Radcliffe contended that elevated testosterone levels "make the competition unequal in a way greater than simple natural talent and dedication." She continued, "The concern remains that their bodies respond in different, stronger ways to training and racing than women with normal testosterone levels, and that this renders the competition fundamentally unfair." However, testosterone is far from the whole story; it's not as if female athletes who win Olympic medals all have naturally high testosterone. One recent study suggested that high testosterone in elite women athletes conferred an average 2 percent advantage for middle-distance runners and a 4 percent advantage for hammer throwers.[16] This is a real effect, but is much smaller than the typical 10 to 12 percent gap in performance between elite male and female athletes in clearly quantifiable sports like running or jumping (as opposed to judged sports like mogul skiing or figure skating).[17]

Chand's appeal was successful. She competed at the 2016 Olympics in Rio without surgery or testosterone-blocking drugs, but she failed to advance past the first round in the women's one hundred meters. Caster Semenya competed as well and took gold in the eight-hundred-meter race. These disparate outcomes support the idea that high natural testosterone is not a uniquely potent special sauce for female athletic success. In its 2015 ruling, the court noted, "While the evidence indicates that higher levels of naturally occurring testosterone may increase athletic performance, the panel is not satisfied that the degree of that advantage is more significant than the advantage derived from the numerous other variables which the parties acknowledge also affect female athletic performance: for example, nutrition, access to specialist training facilities and coaching and other genetic and biological variations."[18]

That last point is particularly notable. Elite athletes, male and female and intersex, often carry rare gene variants that contribute to their athletic performance. A swimmer with an average physique who is just as dedicated to training as Michael Phelps is unlikely to be able to overcome the physical advantages bestowed by Phelps's long limbs and enormous feet. Presently, we haven't identified the gene variants that endow Phelps with his unusual physiology, but they are very likely to contribute strongly to his athletic success.

However, there are some rare cases where making the genetic link to elite sports performance is possible. In the 1960s, the Finnish athlete Eero Mäntyranta dominated Nordic skiing. He won seven medals in three different Winter Olympics. Decades later, genetic testing of his extended family revealed a mutation in the gene coding for the erythropoietin receptor that increased red blood cell growth and survival. As a result, Eero and other affected members of his family had about a 50 percent increase in oxygen-carrying hemoglobin in his blood, a clear advantage in his chosen sport.

Why do we, as a society, easily accept Eero Mäntyranta's genetic advantage in sport as natural talent but argue about Caster Semenya's? It's not that we believe that athletic success should only reflect dedicated effort and not hereditary factors—no one is suggesting, for example, that we ban the tallest basketball players. And it's not that we believe that fairness requires equal access to nutrition or specialized training; we don't see anyone proposing an equalizing handicap for athletes from poor backgrounds in sporting events. The reason that sex categories in sports are unusually divisive is because they occupy a point where complicated biology and deeply held cultural ideas about both sex and fairness collide.

NOT ALL ORGANISMS REPRODUCE sexually.[19] Asexual reproduction by splitting (also called binary fission) is found in bacteria, certain plants, and some invertebrate animals like hydra—a tiny, drifting, freshwater critter distantly related to jellyfish. Why don't more animals reproduce by binary fission?[20] Wouldn't it be faster and easier to dispense with mate finding entirely and just divide and make a genetically identical copy of yourself?[21]

There are two main benefits of sexual reproduction. First, when you inherit two copies of a gene, one from each parent, then the presence of a loss-of-function mutation in one copy is less likely to create a biological problem, as the presence of the other intact copy can usually compensate. Second, and more importantly, the mixing of parental gene variants in each generation of sexually reproducing animals produces more individuality through recombination than can be produced in animals that make exact genetic copies of themselves (clones). Another way to think about this is that genetic diversity in asexual animals can come about only through mutations in DNA, while animals that reproduce sexually have both mutations and the recombination of parental gene variants. Sexual reproduction creates a wider range of genetic variation and thereby forms a broader substrate upon which selective evolutionary pressures can act.[22]

If you're going to reproduce sexually, you need a way of ensuring that two cells from the same individual don't fuse to create offspring—that would defeat the whole advantage of sexual reproduction. So, the specialized reproductive cells, called gametes, must come in two flavors, egg and sperm, and they must be engineered such that egg cannot fuse with egg nor sperm with sperm. This arrangement most commonly requires two different types of organism: males, which make only sperm, and females, which make only eggs.[23] Thus, females and males need specialized egg- or sperm-producing organs: ovaries and testes, respectively. Because sperm are small and motile and fertilized eggs are larger and most often develop inside the female, then males and females will also need other specializations of the reproductive tract: the uterus, vagina, penis, and so on. And, in mammals like us, the females need milk glands to nourish the young.

What's obvious when we look at male and female humans is that, on average, they are different in many ways that are not just directly related to the requirements of copulation, pregnancy, childbirth, and breastfeeding. Adult males are typically taller (figure 10), heavier, leaner, and more muscular, with bigger bones, thicker skulls, and hairier faces. On average, adult females have reduced body hair and higher-pitched voices, with a greater percentage of body fat distributed to breasts, buttocks, and hips. Most of these average differences between males and females are also seen in the fossil record of our hominin ancestors. It's worthwhile emphasizing that we're talking about average differences here.

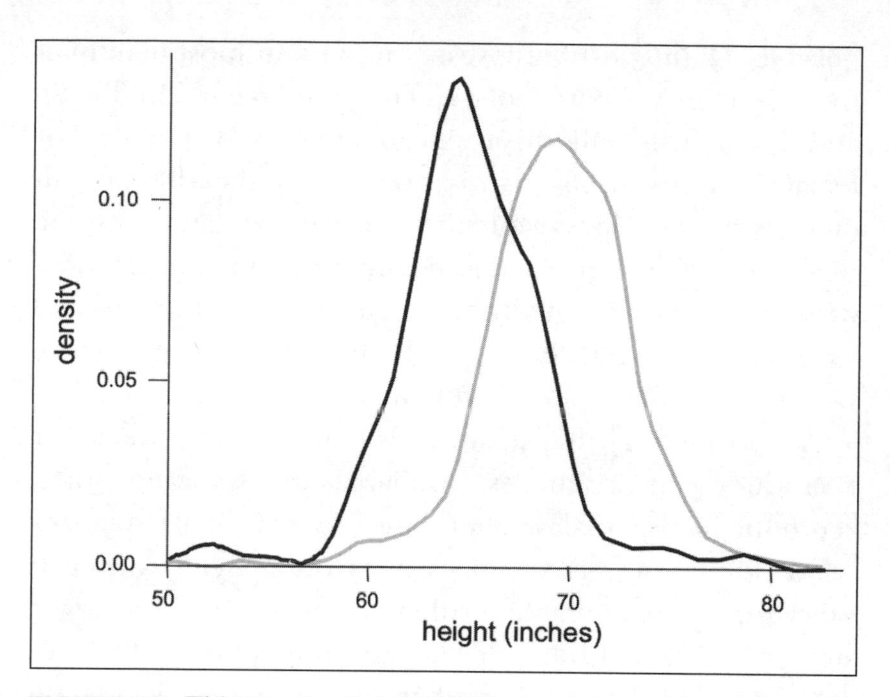

FIGURE 10. This graph of the distribution of adult heights in the United States shows that, on average, men (gray) are taller than women (black). However, there is considerable overlap. Also note that the distribution of height in men is a bit wider, with more men at the extremes and fewer at the average. The degree of overlap between male and female height distributions can be quantified by a statistical measure called standard deviation. The sex difference in height is about 2 standard deviations ($d = 2$).

Of course, there are individual women who are more muscular than the average man and individual men whose voices are higher-pitched than the average woman.

The main explanation for how these physical differences between men and women arose comes from the theory of sexual selection, first proposed by Charles Darwin and since elaborated on and refined by Robert Trivers and many others. It holds that males and females don't mate randomly, but rather seek to mate with those that appear genetically fit, so that their offspring are as healthy and successful as possible. Mating is a big investment, and in most mammals the investment in terms of gametes, pregnancy, childbirth, and child-rearing falls disproportionately on the female. For females, another aspect of this investment is that they are out of the mating business while they are pregnant and for some time after childbirth (such as during breastfeeding), a time when males can potentially mate again. Also, in humans and some other species, the upper bound of the reproductive years is higher for males than females.

All this means that at any given time there are fewer reproductive-age females available for mating than reproductive-age males. There are two main consequences of this difference. First, males must often fight with each other for access to scarce fertile females, hence their larger size and increased bone density and musculature. Second, males must display characteristics that attract females. These may include some of the same traits that allow males to successfully fight or intimidate other males, such as large size or a low-pitched voice. In many species, there are also ornamental male traits that appear to attract females but that are not obviously related to fighting or intimidating other males. The most famous of these, and a favorite of Darwin, was the huge elaborate tail that graces the peacock but not the peahen. Other male signals for sexual selection can involve elaborate behaviors, including calls, gift giving, dances, and nest building.[24] In general, females are much less likely to show such ornaments or behaviors because they face less competition for mates.

Sexual selection theory has been used to explain not just the physical differences between men and women but also both sexual and nonsexual behaviors. It has been claimed that sexual selection has driven men to be promiscuous, risk-taking, aggressive, and violent, while women, with their greater investment in parenthood, are choosy about sexual partners, cooperative, and nurturing. Is this idea true, or is it just a story that has been concocted to reinforce traditional, socially constructed roles for men and women? Whenever there is an explanation that could be used—and indeed has been used—to justify historical and ongoing oppression of women by men, it behooves us to subject it to careful scrutiny.[25]

For sexual selection to operate, it requires that some individuals be more successful in mating and having children than others. If everyone were equally successful at making babies, then there would be no basis to weed out the sexual losers. If sexual selection theory is right, then the variance in reproductive success among males should be greater than that among females, reflecting that males must compete to mate with a limited pool of available, fertile females. The first experimental tests of this idea were performed by geneticist Angus Bateman in the 1940s, using fruit flies, and appeared to show that males do indeed have a greater variation in reproductive success than females. This happens for two reasons. The first is that males show more variation in the number of sexual partners than females. The second is that, when males mate with more partners, they increase the number of offspring produced per additional partner at a higher rate than when females do.

In recent years, Bateman's experiments have been justifiably criticized for flaws in both experimental design and statistical analysis.[26] For some scholars, these critiques are sufficient to toss out the entire edifice of sexual selection. In their view, Bateman's experiments are the basis for sexual selection theory, and so their invalidation destroys the foundation on which subsequent experiments have been built. But this isn't really true. Subsequent experiments and field observations have been inspired by Darwin and Bateman's theory, but they do not rely on Bateman's experimental findings and so should stand or fall on their own merits.

Bateman's predictions have now been tested in a wide variety of animals, ranging from mollusks to insects to fish to mammals. The overall result is that, in the sixty-two different species examined, most but not all males showed higher variance in reproductive success and also more reproductive success as sexual partners were added.[27] Even more informative are the special cases where these results did not hold. For example, species in which the males provide most parental care—like the seahorse or the pipefish (males carry the pregnancy in a special brood pouch, outdoing even the most dedicated Brooklyn hipster dad)[28] and a shore bird called the wattled jacana (males incubate the eggs and raise the chicks)[29]—show stronger sexual selection in females, and the females tend to be larger and more ornamented.[30] These are truly the exceptions that prove the rule. Bateman's experiments may have been flawed, but the Darwin/Trivers/Bateman hypothesis has survived, albeit with certain important caveats.

Another exception that runs counter to the ideas of Darwin, Trivers, and Bateman are species in which females gain reproductive success from mating with multiple males.[31] There are many group-living species in which one or a few dominant females are the only ones to successfully breed. In some cases, like the marmoset or the mole rat, the dominant female physiologically suppresses the reproductive cycles of her subordinates. In others, the subordinate females are allowed to mate but their newborns are then killed by dominant females, as in wild dogs and meerkats. In yet other animals, like savanna baboons, lions, and langur monkeys, most females mate with several males. The anthropologist Sarah Blaffer Hrdy, who first described this behavior in langurs, suggests that such female promiscuity may serve to muddle the paternity of baby langurs, thus reducing the chance that the dominant male will kill newborns that he thinks he did not sire.[32] In recent years, as DNA testing has come into wider use, it has become clear that, in many species that were thought to be monogamous, both males and females are getting a bit on the side. The original Darwinian notion that females are always sexually choosy is incorrect; it holds for many species, but there are more than a few exceptions. Most importantly, the cases where the Darwin/Trivers/Bateman model fails are not random, but fall into particular situations that are well explained by male parental investment, female social structure, infanticide, and a few other factors.[33]

We've just spent some time going over the evidence in critters for and against the hypothesis that sexual selection operates more strongly in males than in females. That is useful to set the stage, but what we really care about are humans. In considering whether sexual selection can explain some of the structural and behavioral differences between men and women, it's worthwhile to note that, among our mammalian cousins, humans are sexually unusual in that we exhibit long-term biparental care, social monogamy, accurate assignment of paternity, and concealed ovulation.

When a human baby is born, its brain is about four hundred cubic centimeters in volume, about the same as that of an adult chimpanzee. It develops at an enormous rate until age five and at a slower rate until about age twenty, when it finally reaches its mature state, with a volume of twelve hundred cubic centimeters. As any mother will tell you, that four hundred cc head just barely fits through the birth canal, and sometimes it doesn't quite make it. In fact, death during childbirth is almost a uniquely human phenomenon.

We need that huge head to house our big brain and be so clever. Why, then, didn't the human birth canal just evolve to be wider, to accommodate big ol' human heads? The most likely reason is that it would require reconfiguring the female pelvis in a way that would interfere with upright posture and so was an evolutionary nonstarter.

Then, the baby is born and is helpless for a really long time. Humans have the longest childhood of any animal. There's no other critter where a ten-year-old cannot reliably make its own way in the world. This means that when the father contributes care, protection, or resources to his children, it's a huge help. By contrast, in most animals, the male is entirely out of the picture, makes no contribution to the care of the offspring, and it's OK.

While many female animals advertise their fertility with swellings, odors, or stereotyped behaviors, human ovulation is mostly (but not entirely) concealed. Despite what perfume manufacturers have told you, we have yet to find any human pheromones, much less those that signal ovulation to males (more on this in chapter 5). This means that most human sex is happening outside of the female's fertile time and is thus recreational, not reproductive. It also means that if a man is to be confident in his paternity, he must monopolize the female sexually throughout her cycle. Not to take all the romance out of the deal, but this is part of the reason why heterosexual marriage (with all of its variants) is such a widespread cross-cultural institution.

This monogamous (or at least serially monogamous) arrangement seems to work out. DNA testing has shown that, contrary to what you might have gleaned from watching *Jerry Springer*, the assignment of paternity to a woman's husband or long-term partner is accurate about 98 percent of the time. This finding holds widely across cultures.[34] False paternity happens, but it is not widespread. It's worthwhile to keep in mind that all of these unusual aspects of the human mating system (long-term biparental care, social monogamy, accurate assignment of paternity, and concealed ovulation) are factors that would appear to reduce sexual selection in men.[35]

By using genealogical records and surveys, we can count people's children and thereby estimate the variance in reproductive success and how it differs between men and women. On average, in a recent meta-analysis of eighteen different human populations across the globe, males did have a larger variation in reproductive success than females, consistent with the Darwin/Trivers/Bateman model.[36] But dig a little deeper into the data and some fascinating details emerge. First, the variance of the effect across populations was large. The ratio of male reproductive success variance to female reproductive success variance was 0.70 in Finns but 4.75 among the Dogon people of Mali. Second, in general, the populations with monogamous marriage systems like those in Finland, Norway, the United States, and Dominica had ratios near 1, indicating no difference in sexual selection in men and women. The populations with polygynous marriage systems (one man, multiple wives), as found among the Dogon, the Aché of Paraguay, and the Yanomami of Venezuela, had much higher ratios, indicating greater sexual selection in men.[37] The most likely conclusion is that sexual selection has indeed operated more strongly in men than in women for much of our human history, and that has given rise to increased height, muscle mass, and physical aggression in males, largely from competition between them.[38]

It may well be that contemporary polygynous societies are a better model of ancestral human social groups than monogamous ones. But that difference is likely eroding as monogamous marriage slowly comes to predominance across the world. That said, even men in most monogamous societies have somewhat higher variation in reproductive success than women, in part because divorced men are more likely than divorced women to remarry and start a second family.[39]

If we look at an affluent, mostly socially monogamous population with access to contraception, such as young adults in the United States, are women, on average, choosier about their sexual partners than men? A study on this topic was conducted by psychologists Russell Clark and Elaine Hatfield.[40] Usually the part of a scientific paper where the methods are described is mind-numbingly dull, but I was amused by Clark and Hatfield's:

> The confederates stood on one of five college quadrangles, and approached members of the opposite sex, who were total strangers. . . . Once a subject was selected, the requestor approached him/her and said: "I have been noticing you around campus. I find you to be very attractive." The confederate then asked subjects one of three questions: "Would you go out with me tonight?" "Would you come over to my apartment tonight" or "Would you go to bed with me tonight?" . . . The requestor carried a notebook which had one of the three requests written on a separate page. The type of request was randomly determined. After the selection of a subject, each requestor flipped a page in the notebook to see what type of request was to be made. . . . Subjects were not approached between class periods or during rainy weather. Subjects were debriefed and thanked for their participation.[41]

This much-cited study was performed at Florida State University in 1978 and was repeated with the same design in 1982. The results of the two studies were almost the same, so I'll report those for the 1978 version: 50 percent of the men and 56 percent of the women agreed to a date and, strikingly, 75 percent of the men and 0 percent of the women agreed to go to bed with the requestor. Yes, that's right. More men were willing to have sex with a stranger than to go on a date. There have since been several replications of the main result in different countries.[42] This study became so well known that the requestor's lines even became the basis for the hit song "Would You . . . ?" by the British electropop group Touch and Go in 1998.[43] One can argue with some details of the experimental design and about the size of the male-female difference, but the overall finding is clear: on average, men are much more willing to have sex with strangers.

Clark and Hatfield suggest that the basis for this striking difference between men and women is that, while women and men are equally interested in sex with strangers, women are holding their desires in check out of a fear of violence, impregnation, or social disapproval. Science historian Cordelia Fine, in her book *Testosterone Rex*, amplifies these points and adds that it may be rational for women to decline stranger sex with men because there is a low chance (11 percent) of them experiencing an orgasm in a casual hookup situation.[44]

So let's do an optimistic thought experiment, in which sexual violence is rare, so-called slut shaming has been banished, and women regularly enjoy orgasms with casual sex partners. Would this situation then reveal that, on average, women and men have equal interest in sex with strangers? We can't really know, but I suspect that men would still be somewhat more interested in casual sex. After all, masturbation is a safe, private, and reliable route to orgasms for both men and women. Yet, even in anonymous surveys, women across the lifespan report less frequent masturbation than men.[45] Among lesbians, the fear of unwanted pregnancy and orgasm-free sex is reduced, but, on average, lesbians report about the same amount of interest in and practice of sex with strangers as straight women (and much less than gay men).[46] There is no question that we have lived and continue to live in patriarchal societies, with the attendant physical and social risks to women engaging in sex with strangers. But the statistics on women's masturbation and lesbian sexual behavior lead me to suspect that, on average, there is a significant biological difference between men and women in casual sexual behavior that has been driven by sexual selection and would remain even if those traditional risks to women were diminished.

There are average differences between the behavior of men and women in the nonsexual realm, but generally the effects are few and of only moderate size. Most measures of personality traits, social interaction, and cognition do not show significant average differences between women and men.[47] Also, as Thekla Morgenroth and her coworkers have noted, we need to be vigilant about cultural assumptions creeping into such assessments. For example, men are culturally stereotyped as being more risk-taking, and this position has been supported by survey data.[48] But if the behaviors we choose to measure to determine risk-taking are already male-associated—like wagering, drug use, and participation in dangerous sports—but do not include female-associated risks—like childbirth (which is, statistically speaking, much riskier than extreme sports) and organ donation (which women do at a higher rate than men)—we've biased the outcome.[49]

Both observational studies in the lab and personality assessments show that men are, on average, more physically and verbally aggressive than women. But the effect is small, about 0.6 standard deviations.[50] Women, on average, appear to be more empathetic (about 0.8 standard deviations), as measured by surveys and observational assessments. Recall that the sex difference in average height was about 2 standard deviations, so aggression and empathy, measured in these ways, are notably smaller effects. However, the assessments by psychologists probably underestimate the real-world sex difference in aggression, as 96 percent of homicides worldwide are committed by men and 78 percent of homicide victims are men (not counting wars).[51] This is unlikely to be due to mere socialization of human males, as a similar sex bias was found in a study of eighteen chimpanzee communities, where 92 percent of killings were committed by males and 73 percent of the victims were males.[52] In the cognitive realm, while there is no significant average difference between women and men in IQ test scores, women tend to do a bit better than men on tests of verbal fluency (0.5 standard deviations) and men do a bit better on tests of spatial perception and mental rotation of objects (0.6 standard deviations).[53]

By far the largest difference between human males and females in nonsexual behavior is found in children's play. Children spend most of their waking hours playing and, on average, boys prefer to play with object toys like trucks, while girls prefer to play with social toys like dolls. And, on average, boys engage in more rough-and-tumble play. These distinctions are present early in life, are broadly cross-cultural, and, unlike most behavioral differences between men and women, they are large. A composite measure of children's play yields a sex difference of about 2.8 standard deviations—even larger than that for adult human height. How do these notable differences in children's play come about? Traditional explanations from developmental psychology hold that children acquire sex-typical patterns of behavior through social learning. Indeed, studies have shown that children tend to pick toys that have been designated for their sex by adults or that they have seen others of their sex choose. From the moment of birth, boys are awash in blue, dinosaurs, and trucks, while girls are inundated with pink and dolls. There's no question that social learning is an important influence on children's play, but is it the whole story?

There are several ways to test the hypothesis that sex differences in children's behavior also have an innate, biological component. One is to look at newborns, who have not yet had the opportunity to be influenced through social learning. In a well-known study by Jennifer Connellan and coworkers, 102 babies (average age = 37 hours) were presented separately with either a real human face (Connellan's) or a mobile consisting of jumbled bits of a photo of her face with a small ball attached, and their reactions were videotaped. The tapes were then cropped to show just the eyes of the baby and were analyzed by independent experimenters who did not know the baby's sex. The result was that boys looked at the mobile more than girls did, and girls looked at the face more than they looked at the mobile.[54] This result has been taken to mean that some behavioral average sex differences are, at least in part, biological in origin. This is an important claim and one that cries out for careful replication.[55]

If the observed differences in children's play behavior have a biological component, then it is likely to result from the exposure of the male nervous system to higher levels of androgens in utero. Indeed, childhood (before puberty) is a time when there are almost no gonadal steroid hormones circulating in the body, so the die must have been cast before then. It is not ethical to manipulate fetal hormones in early development just for the purpose of research, so the fallback has been to study naturally occurring disorders of fetal hormone signaling. Girls with congenital adrenal hyperplasia, who were exposed to elevated concentrations of androgens in utero, show decreased female-typical play and increased male-typical play. Similarly, girls whose mothers were prescribed androgens during pregnancy for medical reasons showed increased male-typical play, including toy choice. Tellingly, the opposite effects on play were produced by fetal exposure to androgen-blocking drugs.[56]

These findings argue strongly for an organizing role for early androgen exposure in determining sex differences in behavior. But we must recall that girls with increased fetal androgen signaling, from either congenital adrenal hyperplasia or maternal androgens, will often have partially masculinized external genitalia. It has been suggested that this appearance will cause the parents to treat these daughters more like sons, thereby affecting their play behavior through social learning.[57] However, observational studies indicate that this is not the case: in fact, if anything, parents tend to encourage female-typical behavior to a greater degree in girls with ambiguous genitalia as a compensatory mechanism.[58]

If normal fetal hormone exposure influences play styles, then we might be able to see sex differences in play behavior in other mammalian species. Indeed, in both rats and rhesus monkeys, males show much more rough-and-tumble play, consistent with preparation for later competition for mates. And in both of these species, treating females with androgens in utero or shortly after birth causes them to play like males.[59] Remarkably, a study of vervet monkeys found sex-typical toy preferences similar to those in human children, with female play appearing to involve preparation for childcare behaviors (figure 11). These preferences are seen even though vervet monkeys lack the social transmission of toy preference found in humans. Indeed, sex-based preferences in the young monkeys were evident with the very first exposure to the toys and without observation of other monkeys interacting with them.[60]

One explanation for these findings, consistent with the vervet monkey results, is that early androgen exposure alters the brain in a way that determines a preference for

FIGURE 11. Reminiscent of human children, female vervet monkeys prefer playing with dolls (left), while male monkeys prefer object toys like cars (right). The female monkey seems to be performing anogenital inspection of the doll, similar to the way vervet mothers inspect their infants. From Alexander and Hines (2002). Used with permission of Elsevier.

male-typical play. However, we also know that boys model their behavior on other boys and men, and that girls model their behavior on other girls and women. So perhaps what androgens are really doing is influencing the brain to attend to and imitate behaviors of males, a form of hormone–social experience interaction.[61]

THERE ARE MANY DISORDERS of the nervous system that have different incidence or severity between males and females. These include diseases that emerge early in life, like autism spectrum disorder, early onset schizophrenia, dyslexia, stuttering, ADHD, and Tourette's syndrome and related tic disorders. Sometimes these effects can be large: autism spectrum disorders are about five times more likely to occur in boys and Tourette's is about three times more likely. Other neuropsychiatric diseases with sex differences typically manifest after puberty and include anorexia nervosa, multiple sclerosis, late-onset schizophrenia, Parkinson's disease, and major depression. Again, sex-related incidence can vary from large (anorexia is about fourteen-fold higher in girls) to moderate (on average, the onset of Parkinson's is about two years later in women). And it can be complicated: multiple sclerosis in about four-fold more common in women but is typically more severe in those men who are affected. This pattern does not allow for a simple explanation, like a multiple-sclerosis-protective factor encoded on the Y chromosome.[62]

Like any trait, we should not assume that the sex differences in neuropsychiatric diseases are all caused by innate biology. It's extremely likely, for example, that the much higher incidence of anorexia among females results, in large part, from societal objectification of women's bodies. And the higher incidence of some adult-onset neuropsychiatric disorders in men, like Parkinson's disease, might relate to higher exposure to environmental toxins in certain industrial workplaces where more men are present.[63]

Similarly, the measures we use to calculate the incidence of some of these diseases depend on the willingness of people to seek treatment and hence be counted in the statistics. Women are treated for major depression at a higher rate than men, but it is unclear if that's because women suffer more depression or because, on average, they are more willing to seek help from a doctor or psychotherapist. And it could easily be that the rate of depression in women (and intersex people) is higher because of ongoing oppression, in much that same way that it is higher among poor people compared to the middle class.[64]

The five-fold higher incidence of autism spectrum disorders in boys and their typical emergence in early childhood has suggested to some, most notably psychologist Simon Baron-Cohen, that exposure to androgens in utero is a significant risk factor for that condition. In his view, informed by experiments measuring testosterone in samples of amniotic fluid and then following those children through early development, autism spectrum disorders are produced when there are unusually high levels of androgens present in utero, resulting in a case of "extreme male brain."[65] At present, however, there has been a notable failure to replicate this basic finding.[66] It may still be that Baron-Cohen's idea is correct, but that testosterone levels measured in amniotic fluid at a single time point are just not a good indicator of developmental androgen exposure.[67] Or it may be that variation in other (*SRY*- and testosterone-independent) genes on the Y chromosome are most important for conferring the increased autism risk in boys. However, to date, the candidate gene variants contributing to autism risk are not located on the Y chromosome.

Together, these results—from sexual behavior to children's play to neuropsychiatric disease susceptibility—indicate that there are some disorders and behaviors that are, on average, significantly different between men and women, and there are a subset of those that appear to have a strong biological component. If this is true, then we would expect to find some important average differences in the function and perhaps the structure of men's and women's brains. The problem that currently limits all human brain research is that we can rarely do invasive studies on living humans.[68] We can take DNA samples from cheek swabs, look in detail at the cellular structure of cadaver brains, and use brain-scanning machines on living people. But these machines are very crude tools. They cannot see individual neurons, nor can they measure the electrical activity of individual neurons or the strength of connections between them. For those crucial experiments, we need lab animals.

In rats, mice, and monkeys, we have solid evidence that, in many different brain regions, there are important sex-related differences in the function of neural circuits. For example, some neurons fire nearly twice as fast in females as in males. Some types of synapses are more easily changed by experience in males. Other neurons have their electrical or chemical properties change over the course of the female estrus cycle in an estrogen-dependent manner. There are more and more of these examples that are accruing as researchers begin to pay attention to sex differences.[69] Importantly, these sex-related differences are not only found in brain regions known to influence sexual behaviors. There are also important differences in the circuits involved in many functions, including motor control, memory, pain, stress, and fear. Sometimes, these differences can be seen in the size of a brain region: There's an area called the medial preoptic area (MPOA) that's involved in sexual behavior and is, on average, larger in males than females. Another region, the anteroventral periventricular nucleus (AVPV), is larger in females, and it's electrically inhibited by a nearby region called the bed nucleus of the stria terminalis (BNST), which, in males, sends ten times more inhibitory nerve fibers to the AVPV. Sorry for the alphabet soup here. The precise names of these regions aren't crucial. The point is that there are many brain circuits that are, on average, functionally different in male and female mammals. What's more, these differences can be changed by experimental manipulation of hormones during development: Males with impaired androgen signaling will have a smaller, female-like MPOA. Females with disrupted estrogen signaling will have a smaller, male-like AVPV. Importantly, while there are many locations in the brain that are, on average, sexually dimorphic in their electrical, chemical, or connection-map properties, there are many others that aren't. These types of experiments are fairly new, and our understanding of sex differences in fine-scale brain function are at an exciting but early stage.[70]

In humans, current technology does not allow us to make these cellular-scale measurements in the intact brain, but there's every reason to believe that the story is similar. For example, the human equivalent of the MPOA is called the third interstitial nucleus of the hypothalamus (INAH3) and, on average, it is also larger in males than in females (in both sexes it's small enough that it can only be measured in human autopsy tissue). In a recent brain-scanning study of 2,838 adults, on average, the gray matter volume of the amygdala (an emotion-processing center) and the hippocampus (a center for memory of facts and events, particularly spatial memory) is slightly larger in males, while that of the prefrontal cortex (involved in self-control and executive function) and the posterior insula is slightly larger in females.[71] An important limitation of these studies is that the brain is plastic, and certain experiences can cause various brain regions to slightly shrink or swell (as we discussed in chapter 3). So, these sex differences in the size of adult human brain regions reflect inborn sex differences convolved with the plastic effects of different experiences of life as a woman or a man. That's why it's particularly useful to study fetal brains, before they are impacted by the long reach of culture. One recent study scanned the brains of 118 fetuses during later pregnancy (twenty-six to thirty-nine weeks gestation) and found significant differences in the resting state connectivity between males and females.[72] This interesting finding awaits replication.

Recently, neuroscientist Daphna Joel and her colleagues posed the question: If there are truly different male and female brains, can we look at a brain scan and accurately determine sex? After all, aside from a small group of individuals with intersex conditions, we can do this easily by looking at the external genitals. To address this question, they examined a large data bank of brain scans of adult women and men. They measured the sizes of many structures and the connections between them and found extensive overlap between the distributions of females and males. In addition, they found that most brains are composed of unique mosaics of features, some more common in females than males, some more common in males than females, and some equally common in both. They concluded that "human brains do not belong to one of two distinct categories: male brain/female brain."[73]

In my view, there are several problems with this conclusion. First, it's a mistake to make a fundamental statement about brains based on an imaging technology with such poor resolution. Their conclusion was not, "With this limited view afforded by present-day brain scanners, we cannot accurately categorize human brains as male or female." Rather, it was a definitive statement about the underlying biology. Second, as has been pointed out by others, the failure of Joel and coworkers to distinguish male and female brains from scans was due to an underpowered statistical design with insufficient comparisons of the various brain measurements. When Adam Chekroud and his colleagues attacked a similar data set of brain scans and applied appropriate multivariate statistics, they were able to classify scans as male or female with 93 percent accuracy.[74] Kevin Mitchell has pointed out that this problem is akin to face recognition.[75] If we take any particular characteristic of a face—the size or shape of the nose or the bushiness of the eyebrows—one will not be able to sort male faces from female ones. Even several of these measurements combined might not be sufficient to make an accurate determination. Yet, when we look at a human face, taking into account many different parameters, we can ascertain sex quite well—similar to the 93 percent accuracy of Chekroud's method. It's not as accurate as a peek in the underpants, but it's quite close.

Third, and perhaps most importantly, I find the whole construction of the question unhelpful. It turns out that we can determine sex reasonably well from an individual's brain scan (and some future high-resolution brain scan will undoubtedly be even better). But, even if we couldn't, so what? The important point is that there are some true *average* sex differences in behavior and brain function between men and women, and a subset of those are likely to be biologically influenced and hence subject to evolutionary forces. The accuracy with which we can assign sex based on an individual's brain scan using present or future imaging technology is unimportant to the larger issue of average sex differences in the human brain.

SO FAR, WE'VE BEEN talking about biological sex as determined by sex chromosomes, variations in hormone signaling, and the random nature of development: female, male, or intersex. Now, let's switch to talking about gender. According to the World Health Organization, "Gender refers to the socially constructed roles, behaviors, activities, and attributes that a given society considers appropriate for men and women." Being socially constructed, gender identity will vary across cultures and times—maleness means something different in modern Japan than it did in Spain during the Middle Ages. While sex is a biological phenomenon that doesn't always sort easily into male and female, gender is even more variable.[76]

Most people are cisgender, meaning that their biological sex and their gender identity match. Recall that, on average, men and women differ in height by about 2 standard deviations. By contrast, men and women differ in gender identity by about 12 standard deviations. That's another way of saying that most people identify with the biological sex they were assigned at birth. But for about 0.6 percent of adults in the United States (about 1 in 167),[77] it's more complicated; this situation is called transgender. Some transgender people feel themselves to be the opposite of their sex assigned at birth. Others feel no particular or lasting association with any gender and thereby identify as nonbinary, agender, gender-fluid, or one of the many other terms (including about seventy now available in the gender box on the Facebook registration page).

If one's gender identity doesn't match one's biological sex, this often, but not always, produces feelings of gender dysphoria. Most transgender people report feeling gender dysphoric at some point during childhood, although for others it can manifest in puberty or adulthood or not at all. The intensity of gender dysphoria can range from mild to strong (the latter often accompanied by major depression or thoughts of self-harm). Depending on individual inclination, opportunity, and cultural practice, gender dysphoria may motivate its sufferers to seek sexual reassignment surgery or hormone therapy.[78]

Neuroscientist Ben Barres, in a letter written to me and many other colleagues as he began his transition to male in the mid-1990s, described his own experience with gender dysphoria:[79]

> Ever since I was a few years old I have had profound feelings that I was born the wrong sex. As a child I played with boy's toys and boys nearly exclusively. As a teenager I could not wear dresses, shave, wear jewelry, makeup or anything remotely feminine without extreme discomfort. I watched amazed as all of these things came easily to my sisters. Instead, I wanted to wear male clothing, be in the Boy Scouts, do shop, play sports with the guys, do auto mechanics and so forth. . . . It is not that I wish I were male, rather, I feel that I already am.

Because these terms can be confusing, I think it's important to make an explicit point: about 0.03 percent of people are intersex, but about 0.6 percent of adults identify as transgender. This means that about 95 percent of people who identify as transgender have normal external and internal genitals. However, they feel that those genitals are not a match (or a consistent match) to their gender identity assigned at birth. So, if most transgender people have sex-typical genitals, how does gender dysphoria come about?

One possible clue comes from the converse statistic: although only about 5 percent of transgender people are intersex, intersex people are unusually likely to switch gender identity at some point in their lives. Recall the condition called 5-alpha reductase deficiency, in which XY people are unable to produce the key fetal masculinizing signal dihydrotestosterone, leading some of them to develop female-typical external genitals and be raised as girls, only later to undergo male-typical puberty. Remarkably, among those raised as girls, most (seventeen out of eighteen in one study) will change to live as men after male-typical puberty. It's not absolute, however. For example, two affected siblings who carry the same mutation, and hence the same impairment of enzyme function, have chosen to live as different sexes, one as a man and the other as a woman.[80]

Since most intersex conditions result from alterations in steroid hormone signaling, then perhaps gender dysphoria results, at least in part, from some alteration in these processes within the brain. One possibility is that there are alterations in fetal steroid hormone signaling throughout the body that are below the threshold necessary to alter the genitals, either internal or external, but are sufficient to change brain circuits that influence gender identity. Another possibility is that steroid hormones are produced in the fetal or newborn brain and can have local effects there. Indeed, estradiol (a particular form of estrogen) can be locally synthesized in some particular brain regions and has been shown to act locally, but brain-derived estradiol has little or no effect in the rest of thc body.[81]

There is some evidence from twin and sibling studies that gender dysphoria has a heritable component, with one estimate yielding 62 percent of variance attributed to genes.[82] This number should be taken as a rough approximation, however, as the sample size was small relative to the incidence of gender dysphoria. To date, there is no convincing evidence implicating variation in any particular gene in gender dysphoria. As with most behavioral traits, the heritable component of gender dysphoria likely results from variation in many different genes acting together or in particular combinations.

Investigating the potential neural basis of gender dysphoria is challenging. It's not amenable to study in laboratory animals. And many people with gender dysphoria who volunteer to be studied in the lab have already been taking hormone therapy or have had surgery, so it's unclear if differences in their brain scans, for example, are a result of such treatment or predate it. One laboratory has suggested that gender dysphoria might be related to the size of the BNST brain region. In adult males, the BNST region is enlarged. Studying a small sample of autopsy tissue, researchers found that the BNST tended to be smaller in male-to-female transgender people compared to cisgender men.[83] However, one problem with such a theory is the fact that the difference in BNST size between males and females seems not to emerge until adulthood, and most transgender people report experiencing gender dysphoria in childhood.[84] Furthermore, these interesting findings have yet to be replicated by another lab. There are also a number of brain-scanning studies examining transgender adults, but the results are inconsistent and the number of subjects is small.[85] In my view, it is likely that there are variants in the function of the brain that contribute to, but do not entirely explain the onset of, gender dysphoria, but, at present, we do not know what they are.

GENDER IS AN OVERWHELMING cultural force that pervades every aspect of human life from the forceps to the stone. And, as we discussed in chapters 2 and 3, we are built to be changed by experience, and so living in a culturally gendered world will inevitably influence our bodies and brains. It's no secret that, even now, in societies that purport to be egalitarian, women, intersex people, and trans people are routinely objectified and denied equality of opportunity. There has been a long history of denying equality of opportunity for women by claiming innate differences in male and female brains and minds. These claims are not just Victorian artifacts. They persist to the present day. So, as a political device, it's appealing to imagine a blank-slate brain, in which differences in the behavior of men and women are solely inscribed by a patriarchal culture. As a lifelong committed feminist, I would be very happy if that were true. But it isn't.

Many claims that have been made, about Martian men and Venusian women and the like, are unsupportable nonsense. In most measures of cognition and personality, men and women are indistinguishable. But, as we have discussed, there remain average biological differences in the brains and behaviors of men and women that are real and significant, and a subset of those are innate. There's much to be learned about innate sex differences in the human brain, and such work must be critiqued and debated and held to the highest standard. But the trend is clear: as we are able to probe at finer and finer scales, to the level of cells, biochemistry, and electrical signals, more average sex differences in brain function are revealed.

Importantly, these differences, whether neural or behavioral, are effects on populations. Even large sex differences, like propensity for physical violence or incidence of autism, do not allow us to make predictions about individuals. There are individual violent women and there are girls on the autism spectrum, even though they are less prevalent than boys. There are men who have multiple sclerosis. While we (and other animals) are built to stereotype and prejudge people based on their sex (as revealed in implicit-bias experiments), we all need to work to eliminate such prejudice against individuals in our thoughts.

Here's what I believe with all my heart. The argument in favor of sex and gender equality, including intersex people and a spectrum of gender identities, must be a moral argument about the way things should be, not a biological argument about the way things are, either in humans or in critters. If tomorrow there were definitive proof of certain inborn average differences in the brain function of women or intersex people or trans people, that would not be an argument for maintaining a system that denies them equal opportunity. The moral argument for equality of opportunity for all can and should accommodate sexual selection theory and certain innate differences between female, male, and intersex people's brains and behavior. Indeed, it is too important a goal to saddle with the ultimately indefensible argument for a blank-slate mind.

FIVE

Who Do You Love?

WHEN I WAS IN HIGH SCHOOL DURING THE 1970S, a popular riddle about a well-liked member of our social group went like this:

Q: Why is Jane bisexual?

A: Because there are only two sexes.

The implication, of course, is that Jane is democratic in her affections and if there were, say, three sexes, then she would be trisexual.[1] I bring up this old joke to highlight that what we have come to call sexual orientation—categorized as gay, straight, or bi—is a crude instrument. If a man had a single early sexual experience with another man, but has since been only romantically and sexually involved with women, do we call him straight or bi? How would he categorize himself, if at all? In 1997, the actress Anne Heche, who had previously only been romantically involved with men, embarked on a well-publicized relationship with the openly lesbian comedian Ellen DeGeneres. That relationship ended after two years, and Heche went on to marry a man. Does that make her presently straight or bi? What about a woman who has been exclusively sexually involved with men but likes to watch feminist lesbian porn? Do these categories even matter?

Is the transgender man who fancies women straight, as befits his gender identity, or gay, as would be indicated by his sex at birth? I'd say straight, but others might argue. To get around this problem, some researchers have employed the terms androphilic (man loving), gynephilic (woman loving), and ambiphilic (woman and man loving) to refer to the object of one's desires without regard for the biological sex or gender identity of the desirer. This highlights that people are not attracted to "those who share my gender identity" or "those who don't share my gender identity"—they are attracted to men or women or both. When people change their gender identity, they almost never simultaneously experience a change in whom they are attracted to, which would serve to maintain a constant sexual orientation of gay or straight.

For most people, their romantic and sexual interests are aligned, but not for everyone and not always. For example, some men in prison will resort to sex with other men for release but would deny any romantic attraction. And lots of us have sexual fantasies that we have no intention or even desire to act on, no matter how much delight they may provide in the realm of the imagination. Some people experience neither romantic nor sexual attraction, or one without the other. It is legitimate to ask if a terminology of sexual orientation should reflect fantasy, desire, overt behavior, or some combination thereof. The answers to these questions are not obvious or straightforward and have been the subject of much debate.

═══

In the southern lowlands of Papua New Guinea and some neighboring islands, there is a traditional belief that there exist two essential bodily fluids: breast milk and semen. The people who live there hold that while all infants require breast milk to grow, boys cannot become men without consuming semen from adult males. Around age ten, boys are removed from their mothers and placed in a separate boys' house at some distance from the village, where they live for months to years. During this enforced segregation, it is the duty of men, particularly maternal uncles, to deliver their semen to boys in order to help them grow through puberty. In some groups, this is accomplished by the man receiving oral sex from the boy, in others by rubbing the semen over the boy's body, and yet others by the boy receiving anal intercourse. This sexual initiation is obligatory and it must always involve the boy receiving semen from the man (the other way around would impede the boy's growth and maturation).

After a boy's initiation ceremony to manhood, he is expected to have a transition period where he is sexually involved with both sexes, but after a few years, he is encouraged to marry a woman and eventually have sex with her exclusively.[2] This example shows how deeply held cultural ideas can profoundly influence sexual behavior. People in the villages that maintain this sexual practice have no words for or concept of gay, straight, or bi. Describing the development of young males as progressing from gay to bi to straight is a construct that we can impose from outside their culture, but it would make no sense to them, and hence its utility and validity are minimal.

═

These days, some people are offended by the lack of subtlety inherent in the traditional sexual orientation categories of straight, gay, and bi. They may adopt newer terms like pansexual (not limited in sexual choice with regard to biological sex, gender, or gender identity), demisexual (sexual attraction only to people with whom they have an emotional bond), or heteroflexible (minimal homosexual activity in an otherwise primarily heterosexual orientation), or even refuse any single-word descriptor of their sexual or romantic feelings. Psychologist Sari van Anders has proposed to replace sexual orientation with a matrix theory, with identity/orientation/status in one dimension and gender/sex/partner number in another.[3] This is a useful endeavor and should be applauded for its accuracy and inclusion, but, in my view, it's a bit unwieldy for everyday conversation. Here, I'm going to use the terms straight, gay, and bi, with the understanding that they are imperfect tools that do not entirely capture the subtlety, range, and dynamism of human sexual expression.

There have now been several large-scale anonymous surveys of sexual orientation performed using random sampling in the United States and Europe. They indicate that approximately 3 percent of men and 1 percent of women identify as consistently homosexual, about 0.5 percent of men and 1 percent of women as bisexual, and the remainder heterosexual. While the results from these surveys have been fairly consistent, the caveat remains that people do not always answer surveys honestly and so there may be some systematic error in the results. In addition, these surveys have mostly been conducted in higher-income, Christian-majority countries, so the results might be somewhat different in other populations.[4]

═══

ASK A STRAIGHT MAN, "When did you decide to be straight?" and he'll likely answer that it didn't feel like a decision at all, but rather like a deep compulsion that became evident around puberty or before. Gay and bi men will give the same answer: in one survey in the United States, only about 4 percent of all gay and bi men reported that they chose their sexual orientation—the rest felt that they were "born that way."[5] In 2019, presidential candidate Pete Buttigieg reinforced this point: "If me being gay was a choice, it was a choice that was made far, far above my pay grade. And that's the thing I wish the Mike Pences of the world would understand—that if you got a problem with who I am, your problem is not with me. Your quarrel, sir, is with my Creator."[6]

Not surprisingly, the scientific question "Is sexual orientation immutable?" has become a political issue. Many on the political right see homosexuality as a harmful and immoral choice made of free will and hence, in their view, undeserving of civil rights protection. Conversely, most people on the political left have advocated for the rights of gay and bi people on the basis that sexual orientation is an innate and immutable trait, hence requiring civil rights protection. Indeed, in *Obergefell v. Hodges*, the landmark US Supreme Court decision that established a constitutional right to same-sex marriage, the majority wrote: "Psychiatrists and others [have] recognized that sexual orientation is both a normal expression of human sexuality and immutable." The court went on to declare that, because sexual orientation is fixed, gay people are compelled to enter into committed same-sex relationships: "Their immutable nature dictates that same-sex marriage is their only real path to this profound commitment."[7]

While men's self-reports would suggest that sexual orientation is nearly always set early in life and remains constant through adulthood, the situation with women is not so clear. Although many lesbians report that they felt solely and consistently attracted to women starting at an early age, a significant fraction have had a more fluid experience of sexual and romantic interest (like Anne Heche did with Ellen DeGeneres). When psychologist Lisa Diamond interviewed seventy-nine lesbian and bisexual women over a ten-year period, she discovered that about two-thirds of the women changed their stated sexual orientation and about one-third changed two or more times. Importantly, and contrary to some widespread cultural ideas, bisexuality was rarely a transition stage on a journey from straight to lesbian. Diamond found that, in many cases, women who had only experienced attraction to men suddenly found themselves falling in love with and being sexually attracted to a woman, or vice versa.[8] In many cases, their atypical attraction was specific to a particular person. For example, a self-identified lesbian who found herself attracted to one particular man might not then become attracted to men generally. Diamond calls this malleability of attraction "female sexual fluidity."[9]

The existence of sexual fluidity complicates the case for gay and lesbian civil rights that has been made to date. Certainly it questions the immutability argument that forms a central pillar of the *Obergefell* decision on gay marriage. However, I would hold, in agreement with Lisa Diamond and legal scholar Clifford Rosky, that immutability should never have been central to the argument for lesbian and gay civil rights in the first place. People who have experienced their sexual orientation as fixed—whether they are gay, straight, or bi—should not be privileged in their civil rights over those who have experienced their sexual orientation as fluid. The fundamental moral argument for gay and lesbian civil rights should be one of individual liberty, not of immutability. By comparison, in the United States we already have laws in place to protect against religious discrimination, even though the practice of a particular religion is clearly not an immutable trait. And we protect the rights of people who change their religion. Sexual orientation is sometimes fluid, more often in women than men, but that is not a valid rationale for denying civil rights to sexual minorities.

THERE'S A DEEPLY HELD idea that the body doesn't lie. You may seek to project a cool and collected exterior in a stressful situation, but your sweaty armpits and racing heart will give you away. There's a similar idea about men and penile erections. I, as a straight guy, may protest that I'm not sexually excited by watching women's Olympic beach volleyball on TV, but when my wife points to the tent in my pants and says, "You're busted!" I must agree, sheepishly.

In the lab, this embarrassment can be studied with a bit more rigor. Cisgender men's erections are measured using a device called a circumferential penile plethysmograph, which is a fancy term for a broad rubber band with a strain gauge inside. As the penis becomes erect and increases in girth, the strain gauge is activated and the signal is sent to a recording device. Volunteers are fitted with this gizmo and are then asked to watch a video screen and simultaneously report their degree of sexual arousal using a computer mouse. The erotic stimuli are various porn films. As a control, videos of nature or sports are used. One might imagine that the ideal porn film to measure heterosexuality in men would involve heterosexual couples. The problem is that, by definition, such a video will feature sexual activity involving both a man and a woman. Experiments have shown that both exclusively gay and exclusively straight men will have erections in response to hetero porn. By contrast, most straight men neither self-report arousal nor do they become erect to gay male porn videos. Similarly, most gay men neither report arousal nor do they become erect to videos depicting sex between two women. More recently, it has been shown that at least some bisexual men respond to male-male and female-female sex videos with both erections and self-reported arousal.[10] The common finding here is that men—bi, gay, or straight—have strong concordance between their self-reports and their erections. Generally speaking, if men say that something excites them sexually, they will become erect in response to it; if they say that something does not, they won't.[11]

The situation for cisgender women is different. There are several ways to measure sexual responses of the female genitals. The one that has been used most widely is called vaginal photoplethysmography. It uses a tampon-size probe containing a light source and a photocell that measures the color of the light reflected from the vaginal wall. The idea here is that female genital arousal includes the production and secretion of natural vaginal lubricant, which is derived from blood plasma. When women are aroused, increased blood flow to the vaginal wall will change its color as a prelude to the production of vaginal lubricant, and this can be detected by the probe. Female sexual arousal is also accompanied by increased blood flow to the external portion of the genitals, and this can be measured using a technique called vulvar laser speckle imaging or another called vulvar thermal imaging.

In some ways, women's responses are like men's: stimuli that women report to be arousing also caused a vaginal response. The difference is that most women also respond vaginally to at least some sex videos that they report as not arousing. Most straight women show vaginal responses to male-male, female-female, and male-female sex videos, even if they self-report arousal to only some of those stimuli. In one study, most women even had vaginal responses to videos of two bonobos (pygmy chimpanzees) having sex, even though nearly all of them reported not being aroused by such activity.

Overall, women's genital responses and their reported arousal are less concordant than those of men. However, there are two interesting details. First, gay and bi women are more sexually concordant than straight women. In other words, those gay and bi women who report no arousal to male-male or hetero sex videos are less likely to have a vaginal response to them.[12] Second, for all women, the vulvar blood flow measures are a better match to self-reported arousal than vaginal blood flow, with less vulvar blood flow in response to stimuli reported as non-arousing.[13]

Why would most women have vaginal wall blood flow responses to stimuli that they report as non-arousing? One possibility is that women's self-reports are unreliable and that they are actually mentally aroused by sexual situations that they deny.[14] I think that this explanation is unlikely. Keep in mind that the subjects for these experiments are volunteers who know that they will be watching sex videos with a probe inserted in the vagina, and so are self-selected to be sex positive. These women are unlikely to be of the inhibited type who would deny stimuli that they found arousing. A more likely possibility is that, because vaginal wall blood flow produces lubrication, it is an adaptive response to situations where vaginal penetration is rapid or nonconsensual (which may have been somewhat more prevalent in human evolutionary history than it is now). Sex researcher Meredith Chivers has suggested that reflexive vaginal lubrication triggered by a wide array of sexual stimuli would reduce the chance of pain, injury, or infection.[15] This explanation is consistent with the observation that vaginal responses are less concordant with reported arousal than vulvar responses, as blood flow to the vulva is unlikely to be as protective.

It's tempting to draw a connection between women's sexual fluidity and women's discordant vaginal responses. Both phenomena suggest that, on average, women are more open to sexual experiences that run counter to their stated sexual orientation than men are. That said, this may be a case of "true, true, unrelated." The reduced concordance between cognitive and vaginal responses in women, particularly women who identify as straight, and the increased fluidity of women's sexual orientation as compared to men may spring from a common neural origin or evolutionary adaptation, but at present there's no evidence either for or against that hypothesis.

═══

MIGHT SEXUAL ORIENTATION BE determined by early social experience, even before overt sexual feelings emerge?[16] Children raised by single mothers are no more or less likely to be straight than those raised by straight couples. Likewise, being raised by a lesbian couple does not increase or decrease one's chance of being straight.[17] Remarkably, a major meta-analysis of the scientific literature by the American Psychological Association failed to find clear evidence that *any* practices of child-rearing—from religion to discipline to education—influence adult sexual orientation.[18] In this way, sexual orientation is like most personality traits: there is some influence of child-rearing (reflected in the shared environment statistic) on behavior early in life, but little influence remains by early adulthood. If you are raised in a household or community that stigmatizes homosexuality, then you may be less likely to come out as gay or bi, but it doesn't mean that you won't feel same-sex attractions.

Some researchers have claimed that both male and female homosexuality can result from childhood emotional or sexual abuse. My reading of the contentious literature on the topic has led me to believe that evidence for this relationship is flimsy. First, there are arguments about the facts. Some surveys have found a positive association between emotional or sexual abuse and later same-sex attraction. But others have failed to find any statistical relationship, or have found an association in women but not in men, or for sexual abuse but not emotional abuse.[19] When there is an association, it tends to be statistically weak. For example, one study found that a history of sexual abuse increased the prevalence of adult same-sex partners, but only by 1.4 percent. Second, even if we were to assume that there are small statistical associations between childhood abuse and adult homosexuality, that doesn't mean that the former caused the latter. One possibility is that reported differences in childhood maltreatment result not from a higher incidence of abuse but rather from a higher incidence of recall among gay men and lesbians. To my thinking, the most likely explanation for this small association is that nascent homosexuality, as expressed by less gender-typical behavior in childhood, slightly increases the probability of child abuse.

Sigmund Freud famously posited that male homosexuality was caused by distant fathers and close-binding mothers (he didn't have so much to say about female homosexuality). One can critique Freud's conclusion by pointing out that his patients were not a representative sample of gay men; rather, they were a subset sufficiently troubled to have sought psychotherapy. That's true, but there is evidence from broader surveys that, on average, gay men report stronger childhood bonds with their mothers and weaker bonds with their fathers than straight men.[20] But just like the purported association with childhood sexual abuse, the devil is in the details. It's probably not that child-rearing strongly influences sexual orientation. Rather, variation in brain circuits that influence gender-typical behaviors, first evident in young children, affect how parents and other adults respond to them. Let's examine how that might come to pass.

If you're a man with a gay brother, your chance of being gay is about 22 percent (versus 3 percent in the general population). If you're a woman with a lesbian sister, your chance of being a lesbian is about 16 percent (compared to 1 percent in the general population).[21] However, there are no such statistical links for opposite-sex siblings. If a woman has a gay brother, that situation is not associated with an increased probability of her being attracted to women. Similarly, if a man has a lesbian sister, that does not increase his likelihood of being attracted to men. There is no familial co-occurrence for gayness or straightness that binds brothers and sisters. The relevant traits here are not "attracted to the same sex as me" or "attracted to the opposite sex as me," but rather "attracted to women" or "attracted to men."

These statistics tell us that sexual orientation clusters in families but do not tell us why. For example, two brothers share, on average, 50 percent of their gene variants with each other, but they also tend to share a similar upbringing. So, for example, if Freud was right and a close-binding mother causes a boy to become gay, then this effect could be found in brothers raised together. As we've discussed, one way to disentangle genetics from upbringing is to analyze same-sex twins. If sexual orientation had no heritable component, we'd expect that the percentage of twin pairs where both are gay would be roughly the same for fraternal and identical twins. Conversely, if sexual orientation were entirely determined by genes, then every gay identical twin would have a gay twin sibling. The best estimates to date, from a population of 3,826 randomly selected twin pairs in Sweden, indicate that about 20 percent of the variance in sexual orientation in women is determined by genes; in men, it's about 40 percent.[22] Some previous twin studies had produced higher estimates of heritability for both male and female sexual orientation—around 50 percent for both—but those studies used self-selected volunteers (in part through advertisements in the lesbian and gay press and at gay pride events) rather than randomly chosen twins, and it is likely that the design biased the results.

The conclusion is that genetic variation is one factor in determining sexual orientation, but it is far from the whole story and it is a somewhat stronger effect in men than in women. As always, it's important to remind ourselves that these estimates of heritability are measures for populations, not individuals. It may be that there are some individual women and men who carry gene variants such that their sexual orientation is entirely genetically determined, and others for whom their sexual orientation has no genetic contribution whatsoever. As with all behavioral traits, there is no single gene that determines human sexual orientation. Rather, many different genes each appear to make a small contribution, and at present we do not have a useful list of these genes.[23] The lack of a single gene for sexual orientation in humans is not an argument against a heritable contribution to variation in this trait. Recall that height is a very heritable trait, but its heritability is determined by small variations in hundreds of genes.

IF CHILD-REARING HAS LITTLE or no effect on sexual orientation, and gene variants have only a partial effect, then why are some people straight while others are bi or gay? What accounts for the missing variation? For cisgender men, fraternal birth order appears to be one contributing factor. What this means is that, if you are male, having older brothers increases your chances of being attracted to males as an adult. This effect, though small, has been found in many different cultures and geographic regions. Having older sisters, younger sisters, or younger brothers has no effect. This does not appear to be a result of child-rearing, as having a biological older brother increases the chance of same-sex attraction in men even if that brother was raised in a different household. Likewise, having an older adopted brother has no influence.[24]

Interestingly, it has been found that men with older brothers have lower-than-expected birth weights, suggesting that the fraternal birth order effect may operate prenatally. One potential biological explanation for the fraternal birth order effects on both sexual orientation and birth weight is that they reflect a progressive maternal immune response directed against foreign male proteins (presumably those encoded by the Y chromosome). In this view, male proteins (or perhaps entire cells) cross into the maternal circulation, are recognized as foreign, and direct the production of maternal antibodies. During the next pregnancy with a male fetus, these antibodies cross through the placenta and influence development of the fetal body and brain. One fascinating recent study has found that mothers of gay sons, particularly those with older brothers, had significantly higher levels of antibodies directed against a protein called neuroligin-4 Y-linked than did the control samples of women, including mothers of heterosexual sons.[25] This potentially exciting finding awaits replication.

Another promising explanation involves exposure to hormones in utero and early postnatal life. In this view, when female fetuses and babies are exposed to higher levels of testosterone, their brains are partially masculinized and this increases the chance that, later in life, they will become sexually attracted to women. Similarly, when male fetuses and babies are exposed to lower levels of testosterone, their brains become partially feminized, thereby increasing the chance that they will eventually become sexually attracted to men. There is evidence in support of this idea from human conditions that alter steroid hormone signaling. For example, as we have discussed, females with a condition called congenital adrenal hyperplasia have increased testosterone levels in fetal life. Even when these girls are treated with testosterone-blocking drugs starting at birth, their brains appear to have been partially masculinized. About 21 percent of women with congenital adrenal hyperplasia report consistent sexual attraction to women (compared to 1.5 percent in the general female population).[26]

This finding is congruent with some experiments in laboratory animals: when guinea pigs, rats, or sheep receive treatments that boost fetal testosterone signaling, the females grow up to display male-typical sexual behaviors. That is, they mount females and fail to exhibit a posture called lordosis, which encourages males to mount them. Similarly, treatments that attenuate testosterone signaling in developing males reduce male-typical sexual behavior in adult rats and sheep. The caveat to these observations is that we don't always know the meaning of the behaviors. When a female sheep mounts another female, is that expressing sexual interest in females, social aggression, or both? Similarly, when a male rat displays lordosis, is that a sign of sexual interest in males, social submission, or both? Or perhaps these behaviors have a meaning that doesn't even occur to us humans.

Can we use average differences in the brain structure of straight men and straight women to test the hypothesis that the brains of gay men are more likely to be partly feminized and the brains of lesbians are more likely to be partly masculinized? The relevant parts of the brain include a portion of the hypothalamus called INAH3, which is larger in straight men, and a bundle of fibers that connects one side of the brain to the other, the anterior commissure, which is larger in straight women. While there have been some well-publicized reports suggesting that the size of the anterior commissure and INAH3 are more female-like in gay men,[27] there have yet to be clear, independent replications of these findings.[28] This doesn't mean that there aren't important differences in the brains of straight versus gay women and men, but, just as we discussed for differences between the sexes, most of the variation is unlikely to manifest as changes in the size of brain regions—the sort of measurement that can be performed with a brain-scanning machine or from autopsy tissue. As a hypothetical example, when comparing lesbians with straight women, lesbians might have lower expression of a gene coding for a voltage-sensitive potassium channel in certain neurons that help to direct sexual behavior toward females. This would render these neurons more electrically excitable but would not change their shape or the volume of the brain region that contains them.[29]

From the earliest portion of childhood, there are, on average, some behavioral differences between girls and boys. As we've discussed, boys are more likely to engage in rough-and-tumble play and to interact with inanimate object toys, while, on average, girls are more likely to play less aggressively and to choose dolls and animals for their toys.[30] When populations of girls and boys were evaluated for gender-typical behavior and then followed to adulthood, an amazing result emerged. Boys who showed highly female-typical behaviors early on were much more likely to become sexually attracted to men as adults (75 percent, compared with 3.5 percent of the general population) and girls who showed male-typical behaviors were far more likely to become sexually attracted to women as adults (24 percent, compared with 1.5 percent of the general population).[31] However, this isn't a universal result: not all tomboys develop sexual attraction to women and not all effeminate boys become sexually attracted to men.

And, of course, as adults not all lesbians are mannish and not all gay men are effeminate. Yet, these findings are striking and suggest a general explanation: sexual orientation is just one aspect of variation in brain function that produces a set of behaviors that can be more or less gender-typical. For example, gender nonconforming girls are more likely to engage in rough-and-tumble play, are less likely to engage in cooperative social play, and are more likely to become sexually attracted to women as they grow up. The most likely explanation is that some combination of social experience, genes, fetal signals such as circulating hormones and immune molecules, and perhaps other biological factors we do not yet understand influence the relevant brain circuits for gender-typical behavior, of which sexual orientation is just one part of the package.

BEYOND GENDER IDENTITY AND sexual orientation lies the even more difficult question about why we are attracted to one person but not another. One thing that's become clear from the huge data set derived from online dating is that—male or female, trans or nonbinary, gay or straight or bi—what we say are our requirements for a mate rarely turn out to be quite so important in the flesh. A straight woman may say that she needs a tall guy who likes the opera, but she may also be happy with a medium-size guy who hates the opera and would rather go to death metal shows. A gay man looking for an extrovert ginger might well find love with a shy blond instead. While certain traits can truly be deal breakers—these days, political affiliation is a line that many will not cross—we're generally not as good at predicting who we'll fancy as we imagine.

The perfume industry would like you to believe that attraction is all about human pheromones, and they are happy to sell you expensive concoctions designed to make your chosen romantic partners consumed with lust. The term pheromone was coined in 1959 to mean signal molecules that are characteristic of a particular species (or, for example, just the females of a particular species) and that trigger a stereotyped behavior in a conspecific target group (like the reproductive-age males).[32] The first pheromone to be discovered was a sexual attractant produced by female silkworm moths (*Bombyx mori*) called bombykol. A tiny bit of this chemical attracts the amorous attention of male silkworm moths from miles around. Importantly, a dab of bombykol smeared on a twig (at concentrations mimicking those secreted naturally by randy female silkworm moths) will still attract males, even if there is no female moth around, and it will have no effect on any other species, insect or otherwise.[33] This is what defines a pheromone: it is produced by every group member of a species and acts on every member of a target group of that same species. While many pheromones are important for sexual behavior, they can also be used to signal social rank, individual or group territory, or the presence of food or danger.[34] Crucially, pheromones are not individual odors; every member of a species group can deploy them and every member of the target group will respond.[35]

Pheromones are used by many animals, from insects to fish to mammals. Some pheromones diffuse freely in the air or water for long distances, while others are so sticky that they must actually be wiped on the recipient.[36] While the original pheromone, bombykol, was a single chemical, it is now understood that some pheromones are mixtures of a few different chemicals. Certain pheromones, like one in the urine of the male house mouse (*Mus musculus domesticus*) can elicit both rapid stereotyped behaviors, like aggression in other males and sexual attraction in females, and slower developmental effects, like accelerating the onset of puberty in young female house mice.[37]

Our mammalian cousins like goats, mice, and rabbits use pheromones, so there's no obvious reason why humans couldn't get in on the party too. We have perfectly fine and sensitive odor detectors in our noses, which are the main way of receiving pheromones. And we secrete all kinds of odor molecules of many chemical classes, often with the metabolic help of bacteria living on our skin. The production of these odor molecules is different in males and females and, as anyone who has attended middle school can attest, they change at the onset of puberty.

One promising indicator of human pheromones, which received wide attention when it was first published in 1971, involved menstrual synchrony in women living together in a dorm at Wellesley College.[38] The idea was that women living together would signal to each other with pheromones to synchronize their cycles (it's not entirely clear why this would be a good idea). A follow-up experiment in 1998 reported that sweat from the armpits of women in their fertile (late follicular) phase, while producing no conscious perception of odor when smeared on the upper lip of recipient women, nonetheless accelerated the recipients' cycle (by advancing the preovulatory surge of luteinizing hormone), thereby promoting menstrual synchrony.[39] Unfortunately, the candidate chemicals from the armpit secretions of fertile women have yet to be identified, and, more importantly, the entire phenomenon of menstrual synchrony has failed to replicate in several subsequent studies.[40] While the idea of pheromone-based menstrual synchrony is still often mentioned in the press, it is now widely regarded as unproven by biologists.

Another line of human pheromone research was inspired by pig breeding. The hormone androstenone is secreted in the saliva of male pigs, and when detected by sows in heat will cause them to reflexively adopt a sexually recipient posture. It's sold to pig farmers by DuPont as the product Boarmate and is used to determine when the sows are sexually receptive and hence ready to breed. When it was discovered that androstenone and some related compounds—androstenedione in men and estratetraenol in women—were also present in human armpits, this was enough to lead several groups of researchers to embark on a set of underpowered and poorly controlled studies to evaluate them as human pheromones. One involved spraying certain chairs in a waiting room with these compounds to see which were later chosen by women and men. The main problem here is that there was never any solid rationale to examine these particular compounds, rather than the hundreds of others present in human armpits. Despite many studies on these chemicals, there is no convincing evidence for their action as human pheromones.

We can take a hint for human pheromone discovery from the recent description of a pheromone in male mouse urine by Jane Hurst of the University of Liverpool. Male mouse urine is a sexual attractant for female mice, and a particular protein isolated from the urine—cheekily named darcin after Mr. Darcy, the smoldering heartthrob of Jane Austen's *Pride and Prejudice*—can entirely mimic its effects on female mice, suggesting that darcin is a mouse pheromone. One possible objection to this interpretation is that perhaps the female mice in the study, rather than responding instinctively to darcin, have simply learned over time to associate it with attractive male mice. To address that concern, Hurst and her group raised females in a sort of strict girls' boarding school for mice, no contact with males allowed. They showed that even these innocent, male-ignorant mice, when sexually mature, responded with sexual interest to either pure darcin or male mouse urine.[41]

As much as some people might like it, we're unlikely to replicate this kind of experiment in humans. To dissociate learning from stereotyped action in our own species, perhaps the best bet is to look at instinctive behaviors in newborns, who haven't had a chance to learn much yet. When mothers are breastfeeding, their areolae swell and secrete tiny droplets of a non-milk liquid from the bumps that surround the nipple called Montgomery's glands. When this secretion is wiped on the upper lip of three-day-old babies, they purse their lips, thrust their tongue, and root around searching for the nipple (figure 12). Crucially, the areolar goo of one nursing mother can elicit this feeding response in a totally unrelated baby.[42] Like a true pheromone effect, it is not dependent on the baby learning to associate feeding with a mother's individual odor.

To prove that there really is a feeding-response pheromone in human lactating mothers' areolar secretion, a chemical or small group of chemicals must be isolated and shown to fully replicate its effect on babies' feeding behavior at naturally occurring concentrations. Then, ideally, depletion of those chemicals would be shown to remove the newborn's feeding response evoked by the areolar secretion.

What all this means is that, contrary to the best efforts of the perfume industry and many popular magazines, at present, there are no proven examples of human pheromones. The best candidate for a human pheromone is in the aforementioned secretion from lactating nipples that causes newborn babies to feed. Not very sexy, that. Does this mean that, unlike mice or moths or goats, we humans simply don't have pheromones for sexual behavior? Some scientists have said yes, noting that a part of the nose that's

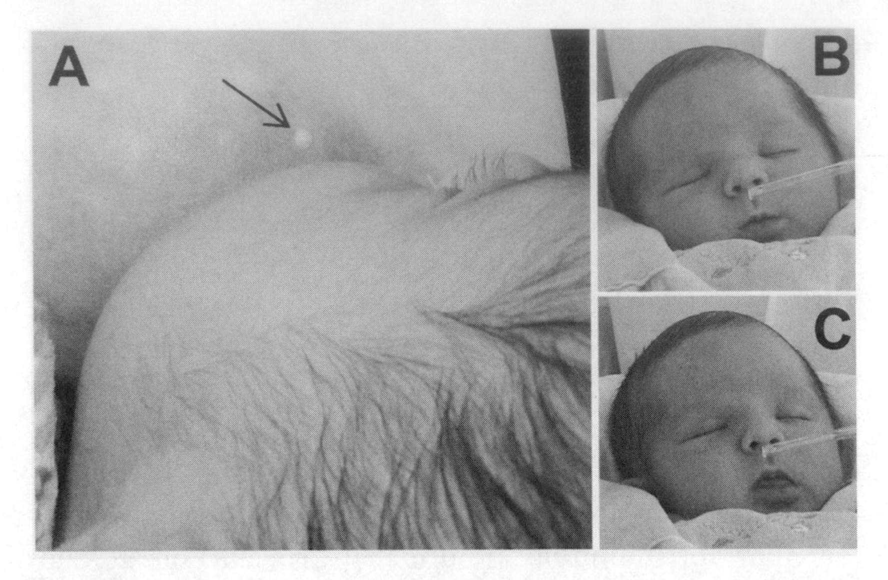

FIGURE 12. Secretion from the areolar Montgomery's gland elicits stereotyped feeding response in a newborn baby. A: Areola of a lactating woman three days after birth with a droplet of Montgomery's gland secretion indicated by the arrow. B and C: Lip pursing and tongue protrusion responses produced by application of the secretion using a clean glass rod. From: Doucet, S., Soussignan, R., Sagot, P., & Schall, B. (2009). The secretion of areolar (Montgomery's) glands from lactating women elicits selective unconditional responses in neonates. *PLoS One, 4,* e7579. Reused here under the terms of a Creative Commons Attribution License.

responsible for pheromone detection in some other mammals, the vomeronasal organ, is merely a vestigial structure in humans (and the other great apes) that's not connected to the brain. But this seems like a hollow argument to me. We know that rabbits and mice can detect some of their own pheromones using the main olfactory system, which we share with other mammals, not the vomeronasal organ.[43] As a result, I wouldn't rule out human pheromones on that basis. More on this topic is to come in chapter 6.

Tristram Wyatt, a pheromone researcher from Oxford University, has pointed out that if we're going to search for human sexual pheromones, a good strategy would be to compare the molecules emitted from men and women before and after the onset of puberty, concentrating on those present solely in the latter.[44] Furthermore, he suggests that we may be looking at the wrong places on the body. Wyatt reminds us of a fact we touched on in chapter 1—that people with the dry-earwax variant of the *ABCC11* gene, who mostly live in northeast Asia, have reduced apocrine sweat gland secretion and hence armpits with minimal odor, yet they manage to attract mates perfectly well. So, sex pheromones may not derive from apocrine gland secretions, but rather from secretions of sebaceous glands, which are present all over the body but particularly on the scalp, face, chest, and crotch. It may be that the armpits are not the best place to hunt for human sex pheromones, and that more embarrassing locations on the body must be swabbed by intrepid researchers.

═══

JUST BECAUSE THERE IS no evidence to date to support the existence of human pheromones does not mean that individual odor is unimportant to sexual attraction. Recently, it has been claimed that attraction can be influenced by a class of molecule called the major histocompatibility complex (MHC), also known in humans as human leukocyte antigens (HLA). The MHC region of the genome, located in a section of human chromosome 6, directs the expression of a group of proteins that play an important role in immune recognition: they bind foreign protein fragments and present them on the cell surface to a key class of immune cell called T lymphocytes. This is a part of the genome that is particularly variable from person to person. It contains thousands of genes variants and hence even more possible combinations. Because you have two copies of chromosome 6, for any protein coded within the MHC region, you may have two identical copies or two different copies, depending on what you inherited from your parents.

The hypothesis, which is well supported in a range of other animals (fish, birds, mice), is that we tend to choose partners with MHC molecules that differ from our own, and that we can detect emitted MHC molecules through the olfactory system. The proposed reason for this choice is that children born of parents with different variants of MHC molecules will have a sort of hybrid immune system and hence be more resistant to infection, as increased MHC diversity improves the immune response to a wide variety of pathogens.

For this plan to work, humans must be able to detect and distinguish MHC molecules through smelling. To test this idea, olfaction researcher Avery Gilbert and his colleagues performed one of the more amusing experiments in human psychophysics. He obtained lines of genetically engineered mice that differed only in their MHC genes and asked if humans could tell them apart by odor.[45] In Gilbert's own words:

> I had blindfolded people sniff live mice in Tupperware containers with holes cut in the sides. Occasionally a mouse tail would get up someone's nose; this seemed to bother some people more than others. The judges also sniffed tiny test tubes filled with mouse urine or dried fecal pellets. . . . For every odor source the results were clear: untrained humans could distinguish between the mouse strains based on smell alone.[46]

Subsequent studies have shown that mice can return the favor and distinguish among human MHC variants. In 1995, these findings and others led Claus Wedekind and his coworkers to perform an investigation that has since become famous as "the dirty T-shirt experiment." Male students were asked to wear the same T-shirt for two consecutive nights, and the next day female students were asked to rate the odors of six different T-shirts. They scored male body odors as more pleasant when the men associated with those odors had MHC types different than their own.[47]

The dirty T-shirt experiment has been replicated and extended several times. Some researchers have found that men also prefer the body odor of women with MHC variants different from their own,[48] while other studies have failed to see such a difference in odor preference.[49] There's a lively ongoing argument in this corner of science, with people ascribing failures to replicate to shaven versus unshaven armpits and fresh versus frozen T-shirts. To my knowledge, all of these experiments have been done on straight, cisgender men and women, so it's unclear how this plays out for gay and bi and trans folks.

Let's assume for a moment that the dirty T-shirt experiments are correct: that both straight men and straight women prefer the body odor of opposite-sex partners with MHC variants different than their own. That still doesn't mean that they will then go on to have children with them. To determine if that is true, one can look at the DNA sequences of mate pairs and, by analyzing both the maternal and paternal chromosomes, ask if MHC variants reflect random mating with regard to MHC, or if people tend to choose mates with MHC variants different from their own. To date, the largest and best study of this type was performed using 239 mate pairs of Dutch ancestry from the Genome of the Netherlands project. Their finding was clear: the distribution of MHC variants reflected mate choice that was independent of MHCs—neither similar nor different MHCs were favored in the chosen mates. The results were statistically indistinguishable from assigning mate pairs at random.[50] It will be useful to see if this result is replicated and if it holds for other populations.

One interesting possibility is that humans do tend to choose mates with MHC variants different than their own, but only under those conditions where the burden of pathogens is high. In cross-cultural surveys, it has been claimed that people in geographical areas carrying a high prevalence of pathogens value a mate's physical attractiveness more than those living in places with less of a pathogen burden.[51] Perhaps the same is true of MHC variants. It may be that people in the Netherlands, with their temperate climate, good public health measures, and well-known penchant for cleanliness are simply the wrong population to see mating based on MHC difference detected through odor.

═

SEX COLUMNIST DAN SAVAGE is completely right when he writes, "When it comes to human sexuality, variance is the norm." Beyond sexual orientation and even beyond one's choice of particular sexual partner(s), there are whole other levels of individual variation in sexual behavior. We have little understanding of why people like their sex a particular way—fast or slow, hard or soft, involving this orifice or that one. And the development of various frills and kinks is almost certainly learned and somewhat serendipitous. There is no evidence for a genetic contribution to foot fetishism or a delight in leather underwear, BDSM practice or sexual peeing. There is, however, a genetic contribution to personality traits like novelty seeking, risk-taking, and obsession, all of which can manifest in the area of sexual behavior, even if they do not direct preferences for particular sexual behaviors.

We may be able to explain some individual preferences for certain sexual acts based on genetic variation in touch sensations. The fine pattern of innervation of the genitals and other erogenous zones does vary from person to person.[52] To wit, some cisgender women may have more nerve endings (or more of a particular type of nerve ending) then average in the vagina and fewer in the labia minora and clitoris. Some cisgender men may have more in the penile shaft and fewer in the glans and anus. But we don't yet know if these anatomical differences actually cause variation in sexual sensation or preferred sexual behavior. There are also likely to be inborn differences in the nerves and brain regions that underlie sexual sensation that are not structural and hence are not detectable with a medical scanning machine or even with a microscope. Such differences would only be revealed by measuring the electrical or chemical signals in the relevant individual neurons that transduce and process sexual sensations. In this way, the biological contribution to our individual preferences for particular sex acts resides not just in our brains, but also in the nerves that course through our skin and viscera.

SIX

We Are the Anti-Pandas

THE BEST THING ABOUT BEING A PARENT OF YOUNG children is that sometimes they forget that you're there. In their minds, the back seat of the car is a private space where they can talk to their friends without fear of parental eavesdropping.

NATALIE (age eight): What's your favorite food? Pickles are mine! I love pickles!

JACOB (her twin brother): Pickles are disgusting, Natalie. My favorite food is French toast.

NATALIE: I like French toast too, but not as much as pickles. Pickles are the best!

SARAH (Natalie's friend): French toast is gross. It smells like eggy farts! I like mangoes.

JACOB: Yuck! Mangoes are slimy!

NATALIE: No they aren't. You're stupid, Jacob.

JACOB: No, you're the stupid one for liking stupid, salty pickles!

You might imagine that this line of conversation could only be sustained for so long. After all, there's no moral point here. People just like different foods. And you're never going to convince someone to change their liking or disliking of a particular food through argument or insult. Still, a back-seat conversation about food preferences could easily go on for thirty minutes, all the way home from school, regularly punctuated with delighted howls of disgust.

Many years later, when I found myself looking for a match on OkCupid, I was surprised to see that, seeking to convey her unique qualities, nearly every woman on the site spent quite a few words on her food preferences.[1] I understand why people turn to food as a topic when trying to paint a picture of themselves as individuals. Everyone has a distinct set of food preferences and they are easy to articulate. Nonetheless I remember thinking, "Hey, CharmCitySweetie! You like spicy food and hoppy beer but despise mayonnaise, mustard, and runny eggs. OK. But really, so what? Are we truly doomed as a couple if you like Stilton and I prefer cheddar?" As it is inclined to do, my mind began to wander and I started to imagine dating websites for other critters:

HOTGIANTPANDA4U: I like to eat bamboo shoots. Nothing else, really. Just bamboo shoots.

SICHUANPANDAMONIUMGAL: Same here. You wanna eat bamboo shoots and chill?

When it comes to food, humans are the anti-pandas. Giant pandas inhabit a small ecological niche—the cloud forests of southwest China—and they eat only bamboo. Humans have spread all over the globe, from the polar regions to the tropics, and so, as a species, we have come to eat lots of different plants and animals. We have succeeded by being food generalists. As a species, we can't be overly predetermined when it comes to food. We must adapt to local availability through learning, giving rise to lots of individual variation. This is the main reason why food preferences are considered such a mark of human individuality.

We can see this idea come out in everyday language, where the phrase "to have good taste" means to have enviable likes and dislikes, including things that never go in the mouth—clothes, music, books, or what have you. Taste has come to mean individual preference considered generally and not just related to food. This meaning is not just a peculiarity of English. In Spanish, for example, the verb *gustar,* which means "to like," derives from the Latin root *gustare,* meaning "to taste," which also gives us the English words gustatory (relating to the sense of taste) and gusto (with an appetite; enthusiasm).

To understand how the food preferences of individuals become so varied, we'll need to explore the neurobiological basis of flavor sensation. In everyday speech, when we say that something tastes good, we're not just referring to the five basic tastes detected by sensors on the tongue (sweet, salty, sour, bitter, and umami), but rather to a unified sense of flavor that blends smell, taste, and touch and refers the sensation to the mouth. When we say "taste" we mean "flavor," and we tend to use the two terms interchangeably.[2] Here, to avoid confusion, I'll use the word taste in the narrow sense to mean only the five basic tastes, and I'll use the word flavor to mean the blended multisensory experience produced by putting food in the mouth.

Most people who go to a doctor complaining that they have lost their sense of taste—which can result from head trauma, side effects of a drug, an infection, or a few other reasons—have actually lost their sense of smell.[3] If you test them by dropping salty or sour or sweet solutions directly onto their tongue, they perceive these sensations normally, proving that their sense of taste is intact. When you think about it for a moment, this is extraordinary. You would never go to the doctor saying that you couldn't hear and be told that, actually, you were blind. You would never report that you couldn't feel your legs and be told that, really, you were deaf—that you had just confused one sense for another. The fact that deficits in smell are often attributed to the sense of taste underscores the unusual degree to which these senses are combined in our experience.

I know it's obvious, but it bears mentioning anyway. By the time you are tasting something, you've probably already investigated its odor through sniffing and have decided to put it in your mouth. The taste sensors on your tongue are there to help you make the further decision: Should I ingest this item or spit it out? In some cases, this is a life-and-death decision. Not surprisingly, the need to make it is found in nearly all animals, and so taste sensors are evolutionarily ancient, probably originating about five hundred million years ago. The modern sea anemone is similar to some of the very earliest animals with neurons. It has no real mouth to speak of, and no brain, yet it can detect bitter substances that have entered its simple closed-end digestive system and then, using its rudimentary nervous system, send commands to contract the relevant muscles and barf it up. In a way, this ancient behavior is the deep evolutionary origin of our cross-cultural human "yuck face," which includes a tongue thrust to eject unwanted food from the mouth.[4]

In humans, the taste-sensing cells are organized into about ten thousand taste buds, clustered into visible bumps called papillae, which are spread across the tongue. The taste buds also reside in the soft palate and the portion of the upper airway called the pharynx, but are not clustered into papillae in these smooth tissues. Each bud is a basket-shaped cluster of fifty to one hundred cells with a tiny pore at the top, and each individual cell within the taste bud is dedicated to one of the five basic tastes.[5] In this way, there are not separate taste buds for, say, sour or bitter, but rather each taste bud has separate detectors for each of the five basic tastes. Crucially, the electrical signals produced when the individual cells are activated by food or drink appear to be kept mostly separate as they are conveyed to the brain.[6] The sensors on the cell surface that bind to taste molecules are proteins, and so their expression is directed by genes. At present, biologists have identified twenty-five human bitter sensors, two sweet sensors, one salt sensor, one sour sensor, and two umami sensors.[7]

Each of these sensors has a particular job in terms of evaluating food and making the decision to swallow or spit it out. Most bitter-tasting chemicals are made by plants and many, like the caffeine in coffee beans or the isothiocyanates in broccoli, are toxins, designed to protect the plants from bacterial or fungal infections or from predators, mostly insects. To a certain degree, when we taste bitter plants, we're eavesdropping on a conversation that has little to do with us: we're the bystanders in an ongoing chemical war between plants and insects.

Other bitter compounds are produced by bacteria. As a result, for most animals, bitter taste indicates either plant toxins or bacterial infection, and so will trigger food rejection.[8] This is an inborn trait. A newborn baby will reject bitter foods with a tongue-thrusting yuck face the very first time they are encountered. No learning is required. Sour taste is also mostly aversive. A little bit of sour can be nice, but strong sour tastes tend to indicate either fermentation, as in sour milk, or hard-to-digest foods, like unripe fruit. Like bitter taste, newborns arrive with sour taste aversion built in.

Sweet taste is the opposite. We are born already finding sweet taste pleasant. If sugar is placed on a latex nipple, a baby will suck longer and harder than if the nipple is coated with plain water. Sweet taste comes from both natural sugars and, to a lesser degree, from the sugars that are released in the mouth as foods containing carbohydrates are chewed and partially broken down by the enzymes in saliva. Throughout human evolution, it has been mostly adaptive for humans to enjoy and consume sweet and carbohydrate-laden, calorie-dense foods, including breast milk. So it makes sense for sweet taste to be hardwired for pleasure.

Umami is primarily the flavor of the amino acid L-glutamate, which is found in many foods with a "meaty" taste, including beef broth, some fish, mushrooms, parmesan cheese, and tomatoes, as well as many fermented products like soy sauce, miso, and fish sauce. Breast milk also has a lot of umami flavor—about as much as beef broth—and this is likely to be at least part of the reason that we are born with a built-in liking for umami.

Salty taste is slightly more complicated. We are born finding salty foods pleasant, but only up to a point. Both babies and adults enjoy salty foods, but the experience of them becomes unpleasant at very high concentrations. This makes biological sense. We need to keep the sodium concentration in our bodies within a fairly narrow range. Either insufficient dietary salt or too much salt can cause trouble in several organ systems, including the nervous system. This problem of optimal salt consumption appears to have been solved in part by having two different populations of taste receptor cells, one that is activated only by low salt and is connected to brain regions that evoke pleasure, and another that is activated only by much higher salt concentrations and is linked to brain regions that produce aversion. At present, the low-salt sensor, called ENaC, has been identified but the molecular identity of the high-salt sensor remains a mystery.[9]

Here's a question: Why are there at least twenty-five human sensors dedicated to bitter taste but only one for sour taste? It's as if nature is trying to tell us something with the math. The reason is that all sour taste comes from the same simple chemical, H^+, also known as a free proton. A lemon may have a different flavor than vinegar because of additional chemicals in each food, most of which are detected by smelling, but the thing that makes both vinegar and lemons sour are free protons (molecules that donate free protons are called acids). When sour taste receptor cells are built to detect free protons and nothing else, they don't need many types of sensor to do it.[10] Likewise, all salty taste comes from sodium ions, almost all umami taste from L-glutamate (and a few structurally similar molecules like L-aspartate), and all sweet taste from a small number of sugar molecules (fructose, glucose, sucrose, etc.) that share a similar chemical structure. So sweet, salty, and umami taste can also function well with one or two receptors each.

Bitter taste, on the other hand, can originate from thousands of different bitter chemicals that are not structurally related. As a result, there's one sensor called T2R38 that appears to be particularly well suited to detect the bitter chemicals produced by certain classes of bacteria, as well as bitter chemicals, called glucosinolates, found in cruciferous vegetables like broccoli and brussels sprouts.[11] Another bitter sensor, T2R1, is one of several tuned to detect, among other things, chemicals called isohumulones that give hop flowers their bitter taste, to the delight of IPA beer drinkers everywhere (including CharmCitySweetie). Some bitter sensors respond to a broad variety of chemicals and others to a single compound. But the overall theme is clear: we need many bitter receptors in order to detect a large and chemically diverse range of bitter substances that we should avoid.

When a particular taste sensor is no longer required in the life of a species due to a change in diet, then the gene that encodes that sensor can accumulate mutations; eventually, some of those mutations will break the gene entirely such that no functional protein is produced. These broken genes are called "pseudogenes." If we look at certain carnivores—like house cats, lions, tigers, vampire bats, and western clawed frogs—they are unable to sense sweet taste; their sweet sensor gene, *T1R2*, has become a pseudogene, shot full of mutational holes. On the other end of the dietary scale, some herbivores, like the giant panda, do not encounter umami in their strict bamboo diet, and so have lost their ability to taste it. Again, we can see the giant panda umami sensor pseudogene *T1R1* lying in the genome like a rusted-out car, up on the blocks.

Perhaps the strangest case of lost taste sensation is found in whales and dolphins, who evolved from a plant-eating terrestrial ancestor about fifty million years ago, becoming carnivorous aquatic mammals. These marine mammals have not only lost their ability to sense sweet taste but they have also lost sour, bitter, and umami.[12] At first look, this is surprising, as whales and dolphins might still need bitter and umami sensors to avoid toxins and enjoy flesh. One hypothesis to explain this limited gustatory repertoire is that because dolphins and whales swallow their prey whole, they do not need to make a decision about which foods get ingested and which are spat out. When everything goes down the hatch, there's no need for taste sensors to inform such a choice. Interestingly, not all marine mammals have lost their taste sensors. Manatees, which are plant eaters, have kept their functional sweet and bitter sensors, reinforcing the idea that the retention of basic tastes across species is determined by diet.

IN ORDER TO MAKE sweet taste intrinsically pleasant and bitter taste intrinsically unpleasant to newborns, a particular wiring pattern must be laid down. The sweet and bitter taste cells each send their electrical signals to dedicated neurons in a region called the taste ganglia. From the taste ganglia, they course into the brain, passing through three processing stations before contacting neurons in the insular cortex, the region of the brain responsible for identifying tastes.[13] Importantly, throughout this taste pathway from the tongue to the insular cortex, the bitter- and sweet-evoked electrical signals are kept mostly separate. Experiments in mice indicate that axons conveying bitter taste information will make synapses and activate neurons in one patch of insular cortex, while those carrying sweet signals will activate a separate but adjacent patch. This pattern, where different types of sensory information are strictly segregated, is called labeled-line signaling. Artificial electrical activation of the sweet taste cortex will make mice behave as if they are experiencing a sweet taste and lick a water spout over and over. Likewise, artificial electrical activation of the bitter taste cortex will mimic a bitter taste and suppress licking behavior.

However, the pleasant or unpleasant quality of taste is not produced in the insular cortex itself. That requires a further continuation of the labeled lines. The sweet cortex sends axons to activate neurons in the anterior basolateral amygdala, while the bitter cortex mostly projects to the nearby central amygdala. Sure enough, artificial activation of sweet cortex nerve terminals in the anterior basolateral amygdala produces a pleasant sensation in laboratory mice, while activation of bitter cortex nerve terminals in the central amygdala produces an unpleasant sensation. When the neurons of the amygdala are prevented from firing, then the mice can still distinguish sweet from bitter (which requires the insular cortex), but the tastes no longer evoke pleasant or unpleasant responses—sweet and bitter are rendered emotionally neutral.[14] If molecular genetic tricks are used to cross-wire the taste systems in mice so that the bitter taste cells send their information along the sweet pathway and vice versa, then sweet tastes will be perceived as both bitter and unpleasant and bitter tastes will be perceived as both sweet and pleasant.[15] The system is modular, with separate stations in the brain for taste identification and taste-evoked emotional responses. When brain researchers speak of a "hardwired behavioral response" in newborns, this is typically metaphorical language. Most often, we don't really have the neural wiring diagram to explain the innate behavior. But here, we do.

IF HUMAN TASTE RESPONSES are really so hardwired, then why don't we all like and dislike the same foods? One reason is that we all carry individual variations in the genes that encode our taste sensors. While almost no one is completely ageusic (insensitive to taste), careful testing in the laboratory reveals that there is considerable variation in individual response to pure taste-activating chemicals applied directly to the tongue (the experiments are done this way to minimize activation of the sense of smell). For example, testing with L-glutamate droplets in a population of French and American adults revealed that about 10 percent can only sense umami taste weakly, and about 3 percent cannot sense it at all.[16] Sure enough, when we look into the DNA, there are tiny variants in the human umami sensor genes (*T1R1* and *T1R3*) that confer greater or weaker sensitivity to L-glutamate. These probably underlie the individual differences in umami sensitivity.[17] Similar stories have been found for genetic variation in one of the two genes that encodes the sweet sensors.[18] In the case of bitter sensors, where there are at least twenty-five different ones, genetic variation in a specific bitter receptor can give rise to individual variation in sensitivity to the particular chemicals that activate it, but that variation does not generalize to all bitter substances. For example, if you carry a bitter sensor variant that makes you extra sensitive to the bitter taste of caffeine, you will also be extra sensitive to quassia bark extract, which activates the same receptors, but you will not necessarily be extra sensitive to sinigrin, a bitter chemical from black mustard seeds, which activates a different group of receptors.[19]

To this point, we've talked about genetic variation in taste sensors that is specific to one of the five basic tastes. In addition to that, there's another, more general type of heritable variation: the number of taste-bud-bearing fungiform papillae on the surface of the tongue. People with lots of papillae are more sensitive to bitter taste, particularly that produced by an artificial chemical called PROP.[20] They have been called supertasters by researcher Linda Bartoshuk, and they represent about 25 percent of the population. Conversely, another 25 percent of people cannot taste PROP at all and so are called nontasters. While nontasters can taste some other bitter chemicals, their overall sensitivity to a range of bitter chemicals is reduced. The remaining 50 percent of people, simply called tasters, are in the middle: to them, PROP tastes bitter but not extremely so.[21] Supertasters also have a somewhat higher sensitivity to sweet, salty, and umami tastes, as well as oral non-taste stimuli like the burn of chili peppers or alcohol. There is a corresponding reduction in the sensitivity of nontasters to these diverse mouth sensations.[22]

I dislike these terms because they are laden with positive and negative associations. Who wouldn't want to be a supertaster? It sounds so cool, like Superman. On the other hand, being labeled a nontaster seems like an insult—the ultimate bland milquetoast. In fact, most supertasters tend to be picky eaters who shun strong flavors to avoid overstimulation. Often, they are particularly averse to vegetables, which are the source of many bitter chemicals. Enthusiasts of strong flavors, who typically enjoy a broader range of foods, are much more likely to be tasters or nontasters. But even this is not guaranteed, as there can also be an interaction with personality traits. A minority of supertasters who are strongly novelty seeking and risk-taking in all aspects of their lives tend to enjoy strong-tasting foods, even if they find them overwhelming.

All of the genes we've discussed so far have been expressed in the taste receptor cells, but they are almost certainly not the whole story. It's likely that genetic variation in the later stages of the taste-processing pathways can also influence taste experience. In that vein, it's worthwhile to reemphasize that taste identification and emotional taste reactions are processed separately. Although there is no evidence of this yet, we can easily imagine that genetic variation in the taste-activated neurons of the amygdala might give rise to changes in the perceived pleasantness or unpleasantness of a particular taste sensation, without changing one's sensitivity to it or ability to identify it.

In addition to genetic variation, there are changes in taste perception that depend on stage of life. For example, when people are given a set of twelve-ounce glasses with increasing concentrations of sugar dissolved in water and asked to identify the one with the ideal level of sweetness (charmingly called the bliss point), most adults will pick a glass with about ten teaspoons added. It's troubling to me that this sugar concentration is even slightly sweeter than most soft drinks. Children have an even sweeter tooth—they choose an average of eleven teaspoons per glass of water.[23] For babies, it's not really possible to make a sugar solution that's too sweet for them. They will happily lick thick sugar syrups that even an eight-year-old would reject. At present, we don't know if these developmental changes in sweetness perception only represent changes in the sensitivity to sweet taste or also involve the perceived pleasantness of sweet taste. The change could be in the tongue or in the brain or both.

The opposite age trajectory is found for bitter tastes: babies are the most intolerant, followed by children, followed by adults. This is generally the case for all five of the basic tastes: as we age, our sensitivity is gradually reduced. On average, women rate bitter tastes as stronger than men, and they experience a temporary increase in bitter sensitivity in the first trimester of pregnancy.[24] It has been suggested that this transient bitter aversion serves to reduce the chance of maternal toxin ingestion during the critical early stages of pregnancy, but this idea, while plausible, is just speculation.

We are born predisposed to like sweet, umami, and mild salt tastes but to dislike bitter and sour tastes. Yet many adults, and even some children, enjoy bitter and sour foods like broccoli, coffee, sour candy, and yogurt. While genetic variation and age-related changes contribute to individual food preferences, they are not the whole story. When fraternal and identical twins reared apart were asked to fill out a self-report of their diet, the results indicated that only about 30 percent of the variation in adult food preference was heritable. Surprisingly, the remaining variation had almost no contribution from the shared environment, suggesting that by adulthood the learned portion of our food preferences mostly occurs outside of the family.[25] Indeed, other studies have shown that even identical twins raised together develop some divergent food preferences by the time they become adults.[26]

Beyond genetic and life-stage influences on taste sensation, there are two main contributors to individual food preference. The first is learning. As we gradually try new foods, we develop associations, good and bad. That cup of coffee is bitter, but it gives me a nice little buzz. With time, I can come to enjoy its flavor. That yogurt is a bit sour, but I like the feel of it in my mouth. Perhaps it can be added to my list of acceptable foods as well. We are all embedded in social groups with complex ideas about foods and eating, and it is our lifelong job as omnivorous food generalists to figure out what we like to eat and to do so in a social context.

The second major factor in individual food preference is the role of the non-taste aspects of blended flavor sensation: mostly smell, but also sight, hearing, and touch. For example, experimental subjects chewing potato chips rate their freshness and appeal based in part on the mouthfeel and sounds produced as the chip shatters. These sorts of multisensory influences are not limited to the food itself. The laboratory of Charles Spence at the University of Oxford has performed many studies showing that our perceptions of food flavors can be affected by all sorts of factors: the sounds of the restaurant, the color and size of the plate or bowl, the weight of the dining implements, etc. In the case of potato chips, the researchers found that the crinkly rattling sound of the chip bag contributed to the perceived crunchiness of the chip itself. In another study, they showed that white yogurt was perceived as slightly sweeter when eaten with a white spoon compared to a black spoon.[27] These effects tend to be small, on the order of 15 percent, but they are significant and underscore that flavor is truly a multisensory experience and that individual food preferences could potentially be influenced by variation in hearing, sight, or touch sensations.

MOST OF OUR EXPERIENCE of flavor comes from the sense of smell. You may not think either Coke or 7UP has much of an aroma, but without smell, they are indistinguishable sweet fizzy water. A slice of steak is merely salty chewy stuff with a bit of umami mixed in. Lemonade is just sweet and sour water. Not surprisingly, people who have lost their sense of smell (a condition called anosmia) take little pleasure in food and often struggle to eat enough to maintain a healthy weight. In addition, they often report difficulty sleeping, cognitive disruption, and loss of motivation and feelings of social connectedness. Most importantly, anosmia sufferers have a significantly elevated risk of depression and suicidal thoughts. Anosmia is a serious, life-threatening condition that goes well beyond the evaluation and enjoyment of food, as we also rely on smell for social and sexual cues, detecting danger, learning, and even navigation.

In relation to food, the sense of smell supports three main decisions. The first is: Where can I find food? The second: Should I put this food item in my mouth? The third, also informed by the sense of taste: Should I swallow this item that's in my mouth or spit it out? Considering these questions is key if we are to develop an understanding of the human sense of smell and how it varies among individuals. The first and second decisions involve evaluating odor molecules originally located outside of the body. These odorants are inhaled through the nostrils to reach a patch of about twenty million specialized olfactory receptor neurons, located in the upper wall of the nasal passages. This sniffing route is called orthonasal olfaction (figure 13).

However, when food is in the mouth, odor molecules released from it are carried by exhaled breath through a passage at the rear of the roof of the mouth called the nasopharynx and so reach the patch of olfactory receptor neurons through the back door. This smelling through exhalation is called retronasal olfaction and is only found in a subset of mammals, including primates and dogs. Importantly, sniffed and exhaled odorants do not reach the array of olfactory receptor neurons in exactly the same pattern and concentration. This means that if you sniff something, the odor you experience is somewhat different than if you hold it in your mouth and exhale. This is probably why some foods, like certain ripe cheeses, can have an offensive odor when sniffed but have a fine flavor (to which odor is a major contributor) once they are in your mouth.[28]

Let's consider the smell of a ripe tomato. The tomato is composed of thousands of different molecules. Of those, only about 450 will be sufficiently small and volatile to be released into the air, where we can measure and identify them using a machine called a gas chromatograph. Of those that travel on the air, only sixteen will bind to the specialized odor receptors on the olfactory receptor neurons with sufficient affinity to reach the threshold of perception by humans and so contribute to the blended odor we call tomato smell."[29] Even then, it's likely that not all sixteen different odor molecules are necessary to evoke a tomatoey sensation. One could probably make a very convincing artificial tomato smell from a subset of those chemicals. In fact, chemical companies do this all the time. For example, roses emit hundreds of volatile molecules, but just one, phenylethyl alcohol, is sufficient to convey a convincing rose smell. In fact, when asked to sniff two vials, one containing natural rose essence and the other containing only phenylethyl alcohol, most people in modern societies, having been exposed throughout life to artificial rose scent in products like hand soap, will misidentify the pure chemical as the natural product.

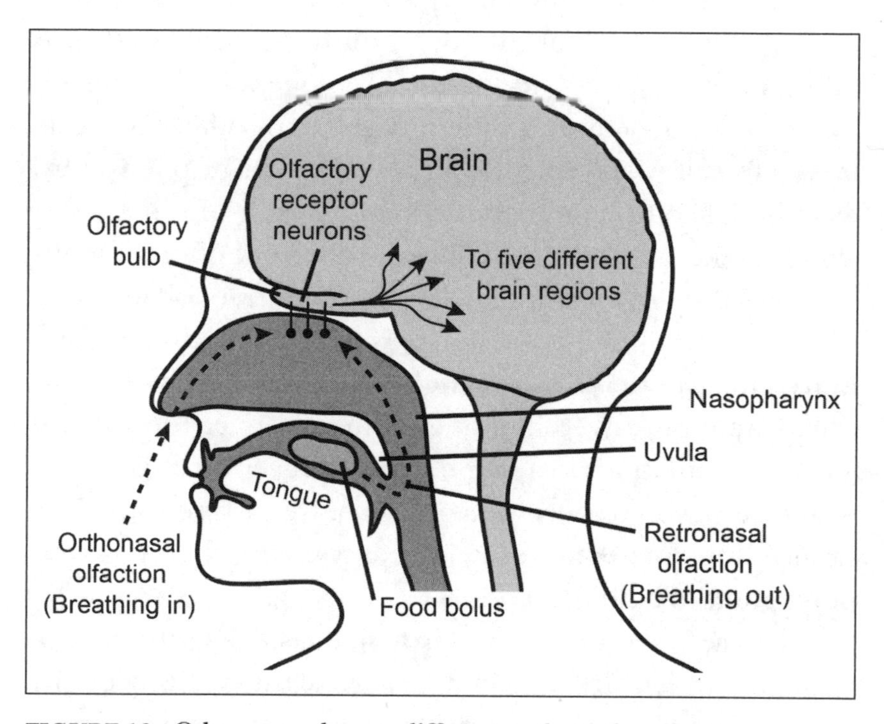

FIGURE 13. Odors can take two different paths to the olfactory receptor neurons. The orthonasal route is engaged by breathing in and samples the external world, while the retronasal route is engaged by breathing out and carries food odors from the mouth to the olfactory receptor neurons via the nasopharynx, the airway at the rear of the roof of the mouth. The olfactory receptor neurons pass their electrical signals through local circuits in the olfactory bulb before these signals branch to reach five different brain regions, each concerned with a different aspect of odor processing. This figure has been adapted from Shepherd, G. M. and Rowe, T. B. (2015). Role of ortho-retronasal olfaction in mammalian cortical evolution. *Journal of Comparative Neurology, 524,* 471–495, with permission of the publisher, John Wiley and Sons. © 2019 Joan M. K. Tycko.

The olfactory receptor neurons are clustered together in a yellowish patch of mucus-covered tissue in the upper wall of the nasal cavity. There are about twenty million of these cells, and each one expresses a single one of the four hundred or so types of olfactory receptor. A particular smelly chemical, like the aforementioned rosy odorant phenylethyl alcohol, will activate many different types of olfactory receptor—perhaps ten to forty of the four hundred possible types. A different pure odorant molecule, like eugenol, which smells like cloves, will activate a different group of olfactory receptors, a few of which might overlap with that of phenylethyl alcohol. Of course, natural smells, like freshly mown grass or wood smoke, are composed of many different odorant molecules at various concentrations, so the pattern of olfactory receptor activation will be even more complex. The main point here is that there is rarely a single receptor for a single odor, even if that odor is from a single pure chemical odorant. If you carry mutations in a single odorant receptor, this usually won't cause you to become supersensitive or insensitive to a single odor, but will more likely have complex effects on your perception of a range of odors.

Like the receptors for a particular taste sensation, such as sweet, which are distributed across the surface of the tongue, the olfactory receptor neurons that express a particular odorant receptor are not clustered together in one patch, but rather are spread across the array of olfactory receptor neurons. However, the information-conveying axons from all of these dispersed receptor neurons converge on one spot in the next processing stage, the olfactory bulb, a specialized part of the brain. This means that different spots in the olfactory bulb where axons converge (called glomeruli) correspond to different types of odorant receptor.[30] Furthermore, at least in mice and rats, the dorsal portion of the olfactory bulb seems to convey signals that result in innate avoidance responses, like the odors of carrion or fox urine, or innate attractive responses, like those evoked by the mouse sex pheromone darcin. From the olfactory bulb, smell information is carried by axons that split to send the information to five different brain regions (figure 13), including the piriform cortex, which is responsible for odor recognition, and the cortical amygdala, which is required to attach positive or negative emotional valence to intrinsically attractive or aversive odors.

I hate to burden you with neuroanatomical details, but there's one subtle point here that's really important for understanding our experience of odors. When the axons from the dorsal olfactory bulb travel to the cortical amygdala, they cluster together, meaning that activation of an odorant receptor can activate both a particular patch of adjacent neurons in the olfactory bulb and a particular patch of adjacent neurons in the cortical amygdala. This is exactly the labeled-line pattern of connection one would expect for innately aversive or attractive odors. However, when the axons from the other parts of the olfactory bulb travel to the piriform cortex, they do not terminate in patches. Rather, they distribute their signals widely across the piriform cortex.[31] This means that a single smell-receptive neuron in the piriform cortex receives information from many, many different types of olfactory receptor, like a giant switchboard.

At first glance, this arrangement seems wasteful. Why go to all of the trouble to cluster information from all of the olfactory receptor neurons together in the olfactory bulb, only to then scatter it pell-mell across the piriform cortex? The likely answer is that the piriform cortex is an olfactory learning machine, where neurons are tuned by the pattern of inputs they have received from the experience of smelling the world. The odor signals to the cortical amygdala are hardwired to produce invariant responses, but the piriform cortex is the proverbial blank slate, waiting to be molded by life experience.

THE POPULAR NOTION THAT humans are inferior to most other mammals in the smell department simply isn't true. There are several factors that determine olfactory ability of a species, including the number of olfactory receptor neurons (about 20 million for us but about 220 million for a bloodhound) and possibly the number of different olfactory receptor proteins (about four hundred in humans, one thousand in dogs, and nine hundred in mice).

While there are some odors that other animals can detect that we cannot smell at all, we're surprisingly good at detecting most odors that originate from plants, bacteria, and fungi. When Matthias Laska of Linköping University in Sweden compared the sensitivity of many species to a panel of different pure odor molecules, he found that, on average, humans were generally more sensitive than many species thought to have a refined sense of smell, including rabbits, pigs, mice, and rats. Dogs, however, kick our human butts, often detecting odors at concentrations over one-million-fold lower than us. When the ability to tell two simple odors apart was tested, humans were in the middle of the pack: worse than dogs, mice, and Asian elephants, but similar to squirrel monkeys and fur seals.[32]

A bloodhound, with its superior ability to detect and discriminate among faint odors, is far better than a human at tracking animals by scent. The other advantage dogs have is their nose location, which makes it easy for them to sample the ground for odors. In one of the more amusing experiments reported in recent years, intrepid smell researcher Noam Sobel and his colleagues showed that college students, fitted with blindfolds, earplugs, and thick gloves, could do a decent job of tracking a scent trail of chocolate extract laid through the grass of an athletic field. All it took was for the students to swallow their pride, get down on all fours, and sniff. The students were pretty good even the first time out and, with repeated practice, could learn to track faster and with greater accuracy.[33]

WHEN WE THINK ABOUT the olfactory abilities of different animals, it's useful to consider the various problems that they have to solve through smell and how these have changed through evolutionary time. Dolphins appear to have no sense of smell at all.[34] This isn't just a result of living in water, however, as salmon, which navigate by smell to find their natal river at spawning time,[35] arguably have the most sensitive smeller of any animal. Mice can sense odors in mouse urine that are undetectable to us but that convey social information to them. Even the very same odorant chemical can have different innate meanings to different animals: the predator odor 2-phenylethylamine is understandably aversive to mice, but it serves as a sex pheromone in tigers. Carrion odors, like the evocatively named molecules putrescine and cadaverine, are aversive to mice but attractive to scavengers like vultures. Each species has a different set of decisions to make based on olfactory information; their sense of smell, from nose to brain, has evolved to support those decisions.

Humans have about four hundred genes encoding functional olfactory receptors and about six hundred more nonfunctional olfactory receptor pseudogenes. Presumably, these olfactory pseudogenes detected odors that were important in the lives of our ancestors, but which are no longer important today. If we compare olfactory receptor genes across species, from the numbers I just mentioned, we can calculate that humans have about 60 percent pseudogenes. Our primate relatives that share three-color vision with us—like chimpanzees, gorillas, and rhesus macaques—have about 30 percent pseudogenes, whereas primates with two-color vision like squirrel monkeys and marmosets have only about 18 percent pseudogenes. This observation has led evolutionary biologist Yoav Gilad and his colleagues to suggest that the development of three-color vision relaxed the evolutionary selective pressure on the sense of smell, allowing the loss of more olfactory receptor genes.[36] For example, one can imagine a scenario in which improved three-color vision could aid in finding ripe fruit that previously had to be sniffed out with a broader range of odor receptors.

On average, humans are good at detecting faint odors, and we're reasonably good at telling two odors apart, but when it comes to identifying even familiar odors by name, we leave a lot to be desired. Imagine if I were to sneak into your home and raid your fridge and bathroom for familiar objects—foods, beverages, cosmetics, medicines—that emit odors. Then, I'd blindfold you and wave the objects under your nose, allowing you to get a good strong sniff. What fraction of the objects do you think you could correctly identify? The answer, from laboratory experiments, is between 20 and 50 percent, with better performance for young adults and a gradual decline with age. These rates of identification fall further if the odors are not necessarily familiar ones.[37] For comparison, if I were to perform a similar task asking you to name familiar objects by sight, you'd likely get nearly 100 percent correct. Vision can reliably trigger the memories for object names, but smell is much less able to do so.[38]

Our mediocre ability to identify odors is likely related to our poor ability to describe them. In most languages, there are only source-based descriptors for various smells: a whisky's bouquet might be described as smoky or peaty. A wine might be redolent of pear, tropical fruit, tobacco, or barnyard aromas. The important point is that all of these odor descriptors refer to a particular source—an object or a process. In English, as in most languages, we do not have abstract terms for odors the way we do for color. A tomato, a firetruck, and a stop sign are all red, and the name of the color makes no explicit reference to the objects that share that property. "Red" is an abstract descriptor, while "smells like a banana" is source based.[39] If we described colors the way we do odors, we'd describe the American flag as having alternating cloud and cherry-colored stripes with cloud-colored stars on a rectangular field of dark sky.

Ethnographers have uncovered examples that challenge the idea that our ability to recognize and name odors is intrinsically limited by brain structure and function.[40] The Serer-Ndut people of Senegal have five abstract odor descriptors. Of these, *pirik* is the smell of bean pods, tomatoes, and various spiritual beings, while *hep* is the smell of raw onions, peanuts, limes, and the Srer-Ndut themselves.[41] Nomadic hunter-gatherers living on the Malay peninsula, who speak Maniq or Jahai, have about fifteen abstract terms for odors. For example, the Jahai word *itpit* can be used to describe odors as distinct as soap, bearcats (*Arctictis binturong*), durian fruit, and certain flowers.[42] The Maniq word *mi?* is used to describe the smell of animal bones, mushrooms, snakes, and human sweat.[43] Perhaps it's not surprising that Jahai speakers perform better than English speakers at naming both familiar and unfamiliar odors.[44] My own reading of the literature leads me to believe that identifying objects by smell and by vision are fundamentally different processes in the brain. Nonetheless, within the constraints of the smell-identification system, intense experience evaluating odors, like that found in populations of hunter-gatherers, can push the boundaries of olfactory naming.

Can we train ourselves to be more like the Jahai? Bianca Bosker was a plucky technology reporter living in New York City with no special affinity for or knowledge of wine or food when she set herself the task of passing the famously difficult Court of Master Sommeliers certification exam within eighteen months. Impressively, she succeeded. She recounts her struggles to rapidly become a wine expert in her hilarious and informative book *Cork Dork.* Bosker writes:

> I liked wine the same way I liked Tibetan hand puppetry or theoretical particle physics, which is to say I had no idea what was going on but was content to smile and nod. It seemed like one of those things that took way more effort than it was worth to understand. . . . I was captivated by these people who had honed the kind of sensory acuity I'd thus far assumed belonged exclusively to bomb-sniffing German shepherds.

In order to progress from a wine novice to a certified sommelier, she had to learn a lot of facts about wineries, grapes, and food pairing, as well as how to recommend and serve wine with aplomb. Most importantly, to pass the exam she had to learn to identify two mystery wines, one red and one white. To accomplish this feat, she had to hone her wine senses by drinking a ton of wine and concentrating hard to isolate the various smell, taste, mouthfeel, and sight components. She learned to attend to many individual sensations, ranging from the color of the wine in the glass, to the degree of burn from the alcohol, to a faint odor of violets that might emanate from an Oregon pinot noir. Most importantly, she had to learn to put words to her wine-drinking experience, and because she's an English speaker, not a Jahai speaker, her words were mostly source based: "Notes of tropical fruit and green grass" and so on.[45] In a sense, she had to train for eighteen months to get some of the olfactory fluency that a Jahai person would have already acquired by age twelve.

It's to be expected that trained wine experts, as well as perfumers and other odor professionals, can perform much better than the average person on recognizing individual familiar odors. However, when familiar odors are blended in a mixture, even the world's most highly trained noses struggle to identify more than three or four components, which is not that much better than untrained people.[46] There appears to be a hard limit on identifying odors in a mixture than can't be extended even with the most intensive training. This makes you wonder what wine experts are reporting when they give tasting notes with lists of ten or more smells.

When experts are evaluating wine, they are using every available sense. What's surprising to me is the degree to which vision can overshadow taste and smell in these evaluations. In one experiment, a panel of wine experts was asked to evaluate two glasses of wine, one a white (a 1996 Bordeaux containing sauvignon blanc and Sémillon grapes) and the other consisting of the very same white wine with an odorless, tasteless organic red dye added to make it appear red. When the panel was asked to describe the flavor of the white wine, they mostly used typical white wine descriptors like grapefruit, pear, and floral bloom. However, when evaluating the white wine dyed red, they almost completely switched to typical red wine descriptors like tobacco, cherry, and pepper.[47] The point here is not to belittle wine experts, but to point out an important aspect of smell: in the real world, olfaction is most often used in combination with other senses, and those other senses can strongly influence the perception of odors.

There's something about the sense of smell that lends itself to imagination and, hence, deception. Consider this devious experiment, reported by E. E. Slosson of the University of Wyoming in 1899:[48]

> I had prepared a bottle filled with distilled water carefully wrapped in cotton and packed in a box. After some other experiments I stated that I wished to see how rapidly an odor would be diffused through the air, and requested that as soon as anyone perceived the odor he should raise his hand. I then unpacked the bottle in the front of the hall, poured the water over the cotton, holding my head away during the operation and started a stop-watch. While awaiting results I explained that I was quite sure that no one in the audience had ever smelled the chemical compound which I had poured out, and expressed the hope that, while they might find the odor strong and peculiar, it would not be too disagreeable to anyone. In fifteen seconds, most of those in the front row had raised their hands, and in forty seconds the 'odor' had spread to the back of the hall, keeping a pretty regular 'wave front' as it passed on. About three-fourths of the audience claimed to perceive the smell, the obstinate minority including more men than the average of the whole. More would probably have succumbed to the suggestion, but at the end of a minute I was obliged to stop the experiment, for some on the front seats were being unpleasantly affected and were about to leave the room.

Lest we think that audiences in 1899 were more susceptible to olfactory suggestion, we can turn to a more recent experiment by psychologist Michael O'Mahony. He arranged for a television show broadcast in the area of Manchester, United Kingdom, to induce olfactory hallucinations in its viewers.[49] At the end of a show about the sense of taste and smell, viewers were told that it was possible to transmit smell by sound. They were shown a bogus apparatus consisting of a two-foot-tall cone called the "taste trap," which had contained a commonly known odorous substance for twenty-three hours, and which was connected to a suitably technical looking nest of cables and electronic equipment with flashing lights. It was falsely explained to viewers that smells were characterized by the frequencies of vibration of the molecules of a substance, and that the vibrations of the molecules causing the odor in the taste trap were being picked up by sensors. A sound of exactly the same frequency as the odor would then be played and broadcast. The brains of the listeners would recognize these frequencies as smell frequencies, which would cause them to experience an odor. Viewers were asked to phone or write to the television company saying whether they had smelled anything or not, and if so what they had smelled. They were particularly urged to write should they not experience any smell at all. Because it was a late-evening show, viewers were told that the smell transmitted would be something they would not normally smell in their house but rather an outdoorsy, country smell. At this point, the studio audience laughed because they guessed it might be manure, so it was clarified that the smell would be a pleasant country smell, not manure. Afterward, 130 people contacted the TV station to report smells. The most common smells were hay and grass, but onions, cabbage, and potatoes also made the list of phantom odors.[50]

It's not just imaginary odors where our perception is malleable. We're deeply influenced by suggestion, context, personal history, and, of course, the evil forces of advertising. If you live in the United States or Europe, then you are likely to have been told by personal care corporations or so-called aromatherapists that the smell of lavender is relaxing and the smell of neroli (an extract from blossoms of the bitter orange tree) is stimulating. This is true, but only if you believe it already. There's nothing intrinsically stimulating about neroli or relaxing about lavender. In one experiment, C. Estelle Campenni and her coworkers exposed college students individually to either lavender essence or neroli essence. The odor was not identified, but in some cases the subjects were told that it was known to be stimulating and, in others, relaxing. Sure enough, students told that lavender was stimulating reported feeling stimulated and had an increased heart rate, while the opposite effects were found when they were told that lavender was relaxing. Of course, the same thing was true of neroli. The odor didn't matter. Only the suggestion.[51]

The power of verbal suggestion on olfactory experience was also tested by psychologists Rachel Herz and Julia von Clef, who presented test subjects with ambiguous odors with labels that carried either positive or negative associations. One of these odors was a mixture of isovaleric acid and butyric acid labeled either "parmesan cheese" or "vomit." Not surprisingly, participants rated the parmesan cheese odor as significantly more pleasant than the vomit odor, even though they were the very same substance. In fact, 83 percent of the subjects were convinced that they had actually smelled two different odors.[52] While all the senses are subject to manipulations by learning, expectation, and context, smell perception seems to be particularly malleable.

Unlike mice, humans are born with few innate emotional responses to odors. This is a good strategy for a wide-ranging omnivore who must learn to consume a variety of smelly foods. Newborns arrive with an inborn aversion to the rotten-fish odor trimethylamine and the spoiled-meat odors putrescine and cadaverine,[53] but an attraction to the secretions of the Montgomery's glands on the breasts of nursing mothers. Presumably, these odors activate the dorsal olfactory bulb to cortical amygdala pathway. But these innate odor responses are the exception, not the rule. Beyond these few limited examples of innate odor responses, our like or dislike of odors is mostly a matter of learning in a social context.

You might think that poop smells are intrinsically unpleasant, and indeed most adults around the world shun fecal odors, but babies happily play with their own waste. They must be taught that fecal odors are disgusting. This teaching is culturally specific. Notably, several groups in Africa, including the Maasai of Kenya and the Mwila of Angola, mix cow dung with other ingredients like butter to make a hair treatment. It probably helps that cows, being herbivores, consume a diet low in cysteine, an essential amino acid that is broken down during digestion to form the intrinsically aversive smelling chemical hydrogen sulfide. Meat eaters consume much more cysteine in their diet, and so their farts and poop are much more redolent of hydrogen sulfide. For this reason, there's probably nowhere in the world where people smear cat poop on themselves.

But what about pepper spray or raw onions or ammonia-laden smelling salts? Aren't those substances universally aversive, even to newborns? The answer is yes, but the reason is because they contain volatile chemicals that are not odors at all. Rather, they activate a special part of the touch sense. In addition to olfactory receptors, the nasal cavity has free nerve endings that are specialized for detecting certain irritating chemicals, like capsaicin (the heat-mimicking chemical of chili peppers, which binds a receptor called TRPV1) and menthol (the cold-mimicking chemical of mint, TRPM8), as well as the warm-mimicking chemicals of horseradish, onions, garlic, and ginger (TRPA1).[54]

Ammonia, found in smelling salts and hydrogen sulfide, as well as rotten eggs and carnivore poop, can also activate TRPA1 receptors.

While these chemical-sensing nerve endings are found in the nose, they are also in the mouth, skin, eyes, and cells that line the airways. This is why, for example, you can feel warmth from chili pepper extract or cooling from crushed mint, even when they are smeared on the skin of your arm. While smell information is conveyed to the brain through the olfactory nerve, the chemical touch sense from the face is carried by a completely different pathway, the trigeminal nerve, and is ultimately processed by different regions of the brain. Strong activation of these chemical touch senses is intrinsically aversive, which is why newborns have an innate, unlearned aversion to chili peppers, raw onions, and ammonia-based smelling salts.

═══

When Leslie Vosshall and her colleagues at Rockefeller University tested 391 adults of varying backgrounds from New York City on odor perception tasks, there were some interesting trends that emerged.[55] Smell acuity was gradually reduced with age. On average, women had somewhat lower thresholds for detecting faint odors than men.[56] Not surprisingly, smokers tended to have reduced olfactory acuity. While you might expect that blind people would have developed a more acute sense of smell by means of sensory compensation, that did not seem to be the case.[57] Overall, individual variability in overall olfactory acuity was the norm. Some people are just better smellers than others, and some people have even lost their sense of smell entirely. However, when you ask people to rate their own sensitivity to smells by merely reflecting on their life experience, most rate themselves as above average.[58]

While there is variation between individuals in overall sensory acuity, there is even more variation in the responses to individual odors. For a given odorant, like 2-ethylfenchol, which has an earthy, mossy smell, certain people might be able to detect it at concentrations one hundred times lower than others, while another group might fail to detect it at all. These differences in sensitivity to particular odors are found widely across individuals. We each live in somewhat different olfactory worlds. My strawberry is not your strawberry and my Gouda cheese is not your Gouda cheese.

When scientists sequenced the genes that encode the four hundred or so functional olfactory receptors in many people, they found that, compared with other parts of the genome, these genes are unusually susceptible to large functional changes—they are more likely to be rendered completely nonfunctional or duplicated in their entirety. In addition, they found an unusually large number of subtler variants in olfactory receptor genes: changes in the DNA that might alter a single amino acid in an olfactory receptor protein and thereby make it more or less sensitive to a given odor. One such study estimated that, between any two unrelated people, on average there would be functional differences in 30 percent of their odorant receptors. That's a lot of variation, and it could account for a significant share of individual odor perception.

The first example of a link between human genetic variation and odor perception involved the chemical androstenone, a metabolite of testosterone, which is perceived by different people as offensive (sweaty, urine-like), pleasant (floral), or odorless. Hiroaki Matsunami and his colleagues found that subtle single-nucleotide variants of a particular odorant receptor called OR7D4 could predict individual sensitivity to androstenone.[59] Since then, subtle genetic variation (single-nucleotide mutations) in other genes coding for odorant receptors have been linked to individual perception of odors. These include the smells of isovaleric acid (cheesy, sweaty), beta-ionone (floral), cis-3-hexen-1-ol (fresh-cut grass), guaiacol (smoky), and several others.[60] While at present we only have a handful of these examples, it's likely that many more will emerge as this topic is investigated further.

One of the best-known individual differences in odor perception involves the smell that some people detect in their urine after eating asparagus. For some, like Marcel Proust, the odor was pleasant. He effused that eating asparagus spears "transform[ed] my humble chamber pot into a bower of aromatic perfume." Benjamin Franklin could also detect an odor in his urine after eating it, but wrote that "a few stems of asparagus eaten shall give our urine a disagreeable odor." While Proust and Franklin disagreed about the pleasantness of asparagus-pee odor, clearly both could smell it. Others are unable to detect the smelly methanethiol and S-methyl thioesters that are produced in urine from asparagus consumption. The rate of asparagus-pee anosmia has been reported to vary across different populations. One recent report that sampled adults of European American ancestry found that about 58 percent of men and 61 percent of women failed to detect asparagus odor in their urine. Because most people don't go around smelling each other's urine, this failure could be a metabolic inability to produce the odorants, an inability to detect them, or both. This question was addressed by some rather unsavory studies showing that people who cannot detect the odor in their own urine also cannot detect it in the urine of known producers, supporting the selective anosmia hypothesis. When the genomes of this population were scanned, three different single-nucleotide mutations were statistically associated with asparagus-pee anosmia, indicating a potential genetic contribution to this trait.[61]

═

Charles Wysocki, a researcher at the Monell Chemical Senses Center in Philadelphia, had assumed that he was one of the 30 percent of people who couldn't smell androstenone. He could weigh it out in the lab and load it into vials for others to sniff, but it never had any kind of an odor to him. But after a few months of working with it, he noticed a new smell in the lab, and sure enough it was the androstenone. Repeated exposure had changed him from a non-smeller to a smeller. Intrigued, he found twenty people who were non-smellers (but who had normal sensitivity to other odors) and had them take a whiff of an androstenone vial three times a day for six weeks. Of those twenty non-smellers, ten became smellers within a week or two. Presumably, the ten non-smellers who didn't improve had broken versions of the key androstenone receptor OR7D4, so they had no possibility of becoming more sensitive. Importantly, the ten androstenone non-smellers who did improve did not have an overall decrease in odor thresholds, as their ability to detect two other smelly molecules, amyl acetate and pyridine, was not changed.[62] This finding has since been replicated by other researchers, and it has been shown that it's not just that non-smellers of androstenone can become smellers, but that people who can smell it weakly can have their sensitivity increased further with repeated exposure.[63]

One possible explanation for this sensitization effect is that intermittent exposure somehow makes the olfactory receptor neurons more responsive to the exposed odor, causing them to send stronger electrical signals on to the odor-evaluating parts of the brain. Another possibility, which is not mutually exclusive with the first, is that the odor-detecting circuits, particularly those in the piriform cortex of the brain, are plastic and change by repeated exposure, so as to efficiently extract the relevant odor-evoked electrical signals coming from the nose. When a tiny electrode is threaded up the nostril to record electrical activity of the olfactory receptor neurons of non-smellers, the androstenone-evoked signals gradually increase with repeated androstenone exposure in those who become smellers, indicating changes in the nose itself.[64] A recent experiment in mice suggests how this might happen. Repeated, intermittent exposure to a particular odorant changed the expression pattern of certain olfactory receptors in the olfactory receptor neurons, possibly rendering the nose more sensitive to that particular smelly chemical in the future.[65]

In a different experiment, androstenone non-smellers were repeatedly exposed to androstenone in one nostril only (using a nose plug/air blower arrangement to carefully restrict exposure to a single nostril both orthonasally and retronasally). After three weeks of exposure, sensitivity to androstenone was found with separate application to either the exposed or nonexposed nostril.[66] This finding suggests that the changes occur in the brain, where information from the two nostrils is combined.[67] This is consistent with a number of brain-scanning experiments showing changes in the electrical activity of odor-processing regions of the brain after odor-related learning.[68]

When Bianca Bosker developed her wine expertise through repeated careful tasting, is it possible that such training increased the sensitivity of her nose to faint wine-related odors? Perhaps there are odorants in wine like androstenone, for which one can increase one's sensitivity with repeated, intermittent exposure. While there is still much work to be done on this question, the early indications are negative. Wine experts (and other odor experts like perfumers) have an increased ability to put names to familiar odors, but their sensitivity to faint odors, even those commonly found in wine, does not appear to be different than the average person off the street.[69]

LEARNING ABOUT FLAVORS STARTS in the womb. The world's expert on this topic is Julie Mennella, who has shown that the substances a mother consumes while pregnant, from foods to cigarettes, will influence her baby's flavor preferences in early life. Odor and taste molecules can pass from the maternal circulation into the amniotic fluid and can be smelled and tasted by the developing fetus during pregnancy. Mennella and her coworkers reported that when pregnant women consumed carrots, anise, or garlic during pregnancy, this increased the acceptance of these flavors when their children were reexposed during infancy or early childhood. The caveats to this finding are that it doesn't necessarily work for every food eaten during pregnancy and that it's not yet clear if this fetal exposure produces lasting effects that influence food choices later in life.[70]

Learning about odors and flavors is a lifelong pursuit for humans. We are not so influenced by the foods of early life that we cannot change our preferences in adolescence or adulthood. We continuously learn to associate particular odors with tastes. This shared experience even makes its way into our language. Most people from the United States will say that vanilla, strawberry, or mint odors smell sweet. On the face of it, this doesn't make sense. Sweet is a taste, not a smell. A substance cannot smell sweet any more than something can sound red. If we break it down, when we say that something smells sweet, what we mean is that, through our experience, we have come to associate that smell with sweet taste. In the case of strawberries, they are naturally sweet when ripe. In the case of caramel, vanilla, and mint, most Americans have experienced these odors in sweet foods like cookies or gum or sweetened drinks. There's nothing intrinsically sweet about these odors—we've just learned to associate them with sweet taste. As a counterexample, in Vietnam, where caramel and mint are used primarily in savory dishes, their odors are not typically described as sweet.[71]

These effects of odor-taste associations can also be studied in the lab. When Richard Stevenson and his colleagues paired vanilla or caramel odor with a sugar solution, it was rated by test subjects as sweeter than the same sugar solution alone, in a population accustomed to vanilla- and caramel-flavored sweets (Australian college students). Similarly, those "sweet" odors reduced the perception of sourness when added to a sour citric acid solution. After novel odors (like water chestnut) were repeatedly paired with sugar solution, they were rated as smelling "sweeter" than they were before the pairing.[72] These experiments reinforce the idea that we are constantly learning (and unlearning) associations between odors and tastes.

We also form a strong association when we eat something and then feel ill afterward. Everyone has one of these stories. For me, I was put off lasagna for about twenty years after a childhood bout of gastrointestinal distress following a family dinner at an Italian restaurant. Obviously, learning strong food aversions is adaptive—if something is likely to have made you ill, you don't want to eat it again, lest it be infected or poisonous.

Compared to other animals, humans are unusually adaptable when it comes to foods, even those that cause a degree of pain. Through learning and deep cultural influences, humans can come to enjoy a wide variety of foods, even those that produce mild pain like chili peppers or raw onions or the ammonia-laden fermented shark dish from Iceland, called *hákarl*. By comparison, it's nearly impossible to train your dog or cat to enjoy chili peppers (please don't try this at home). Even rats, which have a famously varied diet, cannot be trained to like mildly painful foods like chili peppers or wasabi. But humans, as the ultimate food generalists who can live and eat in nearly any location on earth, can learn to overcome our innate aversions to these chemical irritants, as well as to sour and bitter foods and foods with odors that otherwise might indicate dangerous bacterial infection (stinky cheeses, beer, miso, and sauerkraut).

Our individual food preferences are deeply influenced by culture, which, these days, includes advertising. Ethnographers have shown that, around the world, there are specific foods that are a mark of inclusion in nearly every culture. They also often serve as a xenophobic mark of exclusion: "We eat pigs but those other people in the next valley eat fish and it makes them stink." Our individuality, when it comes to food preferences, is not unbounded, but rather is molded and constrained by cultural ideas that influence taste and odor learning.

These cultural ideas are not limited to food odors, and they can be quite specific. One might imagine that the United States and the United Kingdom share a lot of cultural similarities, but there are some notable differences in ideas about odors. One involves the odor of wintergreen (methyl salicylate), which, in a sample of Americans published in 1978, was ranked the most pleasant of twenty-four odors tested.[73] This ranking is in stark opposition to a 1966 survey in the United Kingdom, in which wintergreen was ranked as one of the most unpleasant odors.[74] While there are a few other examples like this, most odors are ranked similarly in the two countries; people in both countries tend to like jasmine but dislike pyridine, which has a stale, fishy smell. The divergent responses to wintergreen do not arise from genetic differences in the odorant receptors between these populations. Rather, it's because of associative learning. In the United States, wintergreen is used in candies and gum. In the United Kingdom (at least in 1966), it was almost entirely used in medicines that are rubbed on the skin for pain relief. The pure sensory experience of wintergreen odor is the same for both groups, but the learned associations, and hence the emotional responses, are completely different.

Cultural ideas about odors can change, and this change is not just an invention of our present trend-chasing society. Pliny the Elder, writing in Rome during the first century CE, opined: "The iris perfume of Corinth was extremely popular for a long time, but afterwards that of Cyzicus. Then vine flower scent made in Cyprus was preferred but afterwards that from Adramyttium, and scent of marjoram made in Cos, but afterwards quince blossom unguent." Being on trend with your Roman perfume was no easy task. Interestingly, these favored perfumes were the same for men and women, a practice that would mostly continue in Europe for centuries. For example, George IV, who ruled England from 1820 to 1830, first encountered a scent on a visiting princess at a royal ball that he later adopted as his own favorite. Fifty years later, styles had changed, and sweet floral blends were deemed exclusively feminine, while men adopted more woodsy scents.[75] Although perfume companies might tell you otherwise, there's nothing intrinsically womanly about the smell of flowers. It's merely a cultural construction of the present moment, aided by an unusually malleable human olfactory system

SEVEN

Sweet Dreams Are Made of This

My parents were plenty hot for each other, but they just couldn't make it work in the domestic sphere. They met in Chicago in the early 1950s, when my mother was a teacher at a school for special-needs kids and my father was a medical student. He was doing a clinical rotation in pediatric neurology and was visiting my mother's school together with the attending physician. They locked eyes across the room, met for a drink later, and were married a few months after that. But domestic bliss did not ensue. Within a year they were divorced, and both had moved from Chicago—she to New York City to work in publishing and he to Los Angeles to do a residency in psychiatry. After a few years apart, he called her on the phone and implored her to give it another try. He must have been persuasive, because she moved to Los Angeles and they were soon married for a second time. And then divorced again.

You might think that's the end of this particular love story. Though twice divorced and living apart, they just couldn't keep their hands off each other. One day when I was twelve, I was in the car with my mom, driving down Sepulveda Boulevard, a gritty, charmless commercial street in West Los Angeles. "You see that hot pillow joint over there?" she said. "That's where you were conceived, back in 1961. In a motel bed with a coin-operated Magic Fingers attachment.[1] Afterward, your father and I joined your Aunt Felice and Uncle Alan for dinner."

That's how I became an atypical child of divorce. My parents were indeed double divorced, but only before I was even conceived. There was no childhood divorce trauma for me. Living with mom on school days and with dad on weekends was all I ever knew. And it was just fine. They were both attentive, loving parents, and I never wished for a more conventional family situation.

I never saw my parents living together, but, based on their habits, I can't imagine a more ill-suited couple. He was messy and she was a neatnik. She loved to cook but he preferred restaurants. She was supersensitive to background noise and he liked to monitor the news on the TV or radio. She didn't even own a TV until 1974, when she bought one to watch Nixon resign. I can't imagine them sharing a household.

I've often wondered if many small differences in lifestyle can accumulate so that a couple just can't live happily together, even if things are otherwise OK. In my limited view, the traits that most separated my parents, and held the greatest potential for discord, had to do with time. My dad was a night owl while my mom was usually in bed by 9 p.m. and rose with the lark at 5 a.m., even on those days when she didn't need to go to work. She was early to every appointment, while he ran notoriously late, annoying her to no end. Now, it may well be that even if their internal clocks had been aligned, they still would not have had a good marriage for any number of reasons. But a boy can wonder.

WE ALL KNOW PEOPLE who are extreme owls or extreme larks, and we know that most people fall somewhere in the middle. However, when sleep surveys of many thousands of people were conducted, some interesting trends emerged. In order to separate people's natural inclinations from the demands of waking for work or school, the time of sleep and waking was calculated for Friday and Saturday nights. The midpoint of sleep on those nights is taken as an indication of how one's activity rhythms relate to dawn and dusk, called a chronotype. Figure 14 shows the results from a survey of 53,689 adults in the United States.[2] The midpoint of sleep ranged from about midnight to 9:30 a.m., with the average at about 3 a.m.

If we break this down, we find, not surprisingly, that high school and college students have the latest average chronotype of any age group. On average, women have slightly earlier chronotypes than men up to age forty, but slightly later chronotypes than men after that point. In both men and women, the variability in chronotype decreases with age; there are fewer extreme larks and extreme owls among older folks. This might be because there are biological factors that influence chronotype and change with aging. It might also be due to aging-related changes in lifestyle, like a reduced burden of childcare. Or, importantly, it might be that extreme larks and extreme owls are somewhat more likely to die young. If you are an extreme lark or an extreme owl, the likelihood that you will be able to find a work schedule that matches your chronotype is reduced, and there is evidence that this mismatch between chronotype and work schedule can have serious health implications. For example, in a large study of nurses, such a mismatch (owls working daytime shifts or larks working nighttime shifts) significantly increased the incidence of type 2 diabetes.[3] Chronotype–work schedule mismatch was also statistically associated with higher incidence of cancer, cardiovascular disease, and stroke.

Chronotype is also influenced by cultural factors. Just ask any American who has walked into a restaurant in Spain at 8 p.m. for dinner and found it empty. In one recent smartphone-app-based worldwide survey, Belgians and Australians had the earliest average bedtimes (around 10:30 p.m.) while Spaniards, Brazilians, Singaporeans, and Italians had the latest ones (around midnight).[4] One might imagine that chronotype and sleep duration would be affected by the use of artificial lighting and heating, not to mention more recent light-emitting diversions like smartphones and computers.

It is widely asserted that the technology of modern life leaves us sleep-deprived, but the evidence for this is not so clear. It should be noted that blaming sleep disturbance on the ills of modern life is not a recent innovation. When Henry David Thoreau retreated to an isolated cabin on Walden Pond in 1845, his main motivation was to alleviate his insomnia, which he blamed on trains and factories (it didn't work).

It order to investigate how people slept before artificial lighting and heating became widespread, historian A. Roger Ekirch studied diaries, books, and accounts of travelers in preindustrial Europe. He has claimed, based on accounts of "first sleep" and "second sleep" in these writings, that "until the modern era, up to an hour of quiet wakefulness midway through the night interrupted the rest of most Western Europeans, not just napping shepherds and slumbering woodsmen. Families rose from their beds to urinate, smoke tobacco, and even visit close neighbors." In Ekirch's view, this "segmented sleep" was the norm, not just in Europe but in preindustrial societies generally, and that the present emphasis on "consolidated sleep" as the ideal form of rest is a recent and unnatural outgrowth of our technological age.[5]

If Ekirch's assertion is correct, and, as he has claimed, it applies to people living in the tropics as well as the temperate latitudes, then one would expect that present-day preindustrial people, who mostly live in the tropics, would show segmented sleep as well.[6] Unfortunately, this has not been borne out in wristband actimetry studies performed among the Hadza of Tanzania, the Tsimané of Bolivia, and the San people of Botswana and Namibia.[7] Consolidated sleep was also found among the Toba/Qom people of Argentina,[8] and in preindustrial *quilombola* dwellers in Brazil.[9] To my knowledge, there are no reports that have found widespread segmented sleep in any population, either pre- or postindustrial. In the absence of such data, I remain deeply skeptical of Ekirch's assertion that our ancestors typically slept in a segmented pattern.

However, beyond the dubious claims of a segmented pattern, are there some other differences in how preindustrial people sleep? There is general agreement that, on average, preindustrial people tend to have their chronotype shifted about an hour earlier, like Belgians compared to Italians. The question of sleep duration is less clear, with some researchers finding about an hour increase in sleep duration among preindustrial peoples,[10] and others, working with different populations, finding no significant difference.[11] It should be emphasized that there are many local influences on sleep duration besides electric lights, cell phones, and heaters (such as noise and social customs), and these could contribute to the variability of sleep-duration results in these studies.

In recent years, it has become popular for certain hard-charging leaders of government and industry to proudly assert

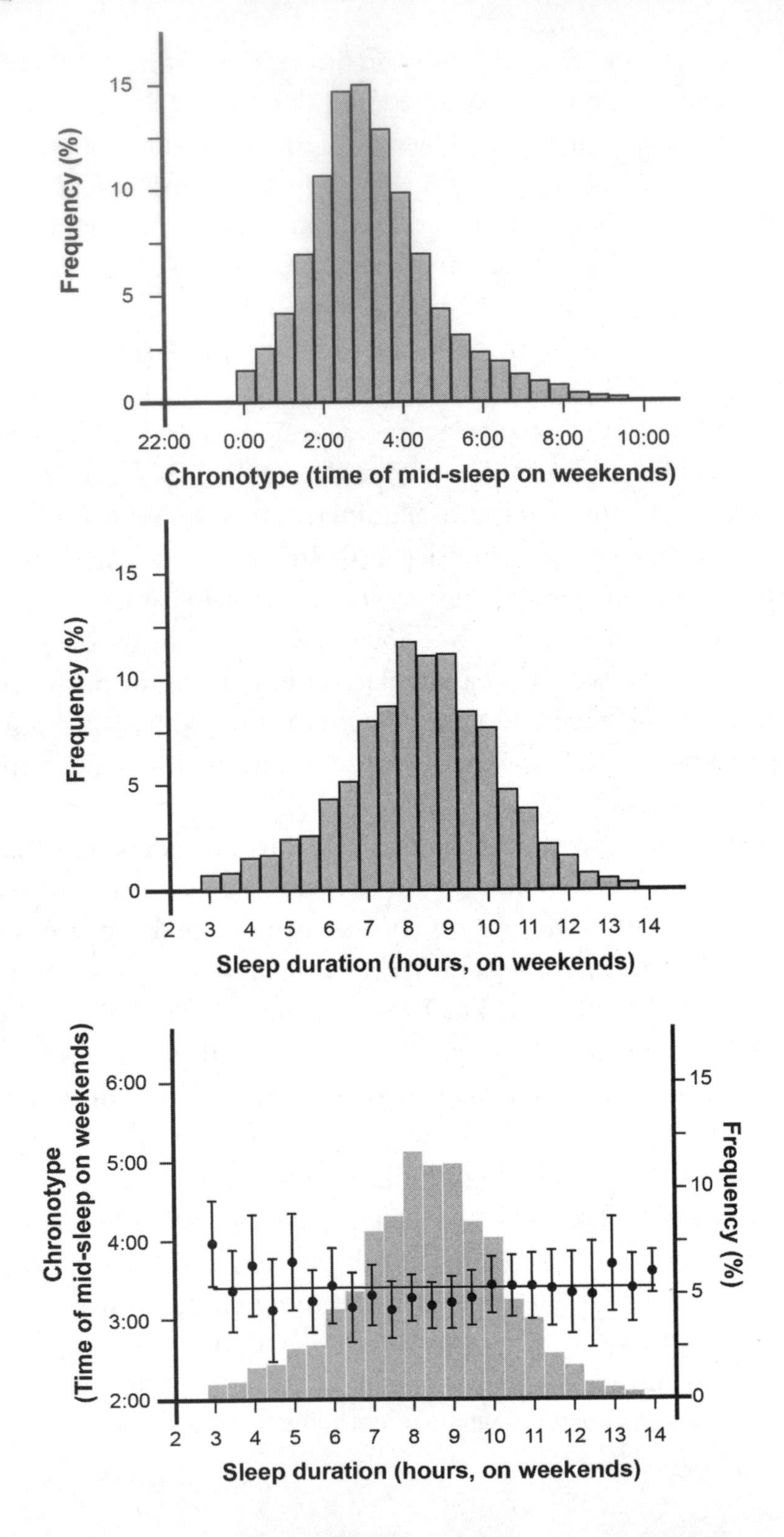

Frequency (%)
15
10
5
0
22:00
0:00
2:00
4:00
6:00
8:00
10:00
Chronotype (time of mid-sleep on weekends)
Frequency (%)
15
10
5
0
2
3
4
5
6
7
8
9
10
11
12
13
14
Sleep duration (hours, on weekends)
Chronotype
(Time of mid-sleep on weekends)
6:00
5:00
4:00
3:00
2:00
Frequency (%)
15
10
5
0
2
3
4
5
6
7
8
9
10
11
12
13
14
Sleep duration (hours, on weekends)

that they need but a few hours of sleep every night. President Donald Trump, Prime Minister Margaret Thatcher, Tesla founder Elon Musk, and fashion designer Tom Ford have all claimed to get by on four hours or less. Those claims may be true, but, if so, they are very rare cases. Figure 14 shows that individual sleep duration ranges widely, from three to fourteen hours, with an average of about 8.5 hours. Only a small percentage of adults in the United States report four hours or fewer. Interestingly, there is no significant correlation between chronotype and sleep duration. You are just as likely to need a lot of sleep if you are a lark as if you are an owl. This lack of correlation suggests that chronotype and sleep duration are under mostly separate control in the brain.

IT'S NOT JUST A human thing. Animals, bacteria, and fungi all match their biological rhythms to the solar cycle of day and night. Even many plants open and close their flowers at particular times of the day. Daylight provides energy to drive photosynthesis in plants, and warmth and light to see by, but those daytime photons can also damage DNA. Because DNA replication, which is part of cell division, renders DNA particularly sensitive to damage by light, nighttime is favored for cell division and cell repair. In humans, it's not just sleeping and waking that are keyed to the solar cycle, but also body temperature, eating, digestion, mental focus,

---

FIGURE 14. Results from a recent sleep survey of adults in the United States. The top panel shows the distribution of chronotypes, measured as the midpoint of sleep on weekends. The middle panel shows the distribution of sleep duration and the bottom panel shows chronotype and mean sleep duration superimposed, illustrating that these two aspects of sleep are uncorrelated. Adapted from Fischer et al. (2017). Used with permission of a Creative Commons Attribution License (CC BY). © 2019 Joan M. K. Tycko.

hormone secretion, growth, emotional state, and many other functions (figure 15). Even our most tender sentiments are influenced by the solar cycle—the most popular time for sex is reported to be 10 p.m.[12]

Do daily cycles of activity require a twenty-four-hour clock within the body, or are these behavioral and physiological rhythms solely driven by external cues like sunlight and ambient temperature? If you were to go live in a dark cave of constant temperature, without a watch (or Wi-Fi), your daily rhythms of sleeping and waking, body temperature, etc. would persist, but this cycle would become gradually desynchronized from the clock of the external world. For each twenty-four-hour period spent in the cave, your bedtime would shift about twenty minutes later. Similarly, if we were to take cells from your skin or liver and grow them in the dark in a culture dish filled with nutrient fluids, they would also show rough daily rhythms in their metabolic activity and in the expression of certain genes. These findings demonstrate that there is indeed an internal clock distributed throughout the body, but that it requires information from the outside world to remain synchronized with the solar cycle. Because your internal timekeeper runs approximately, but not exactly, twenty-four hours long, it's called

---

FIGURE 15. We contain biological clocks that serve to entrain our bodies to the daily cycle of darkness and light. The twenty-four-hour-long revolution of the earth imposes day and night cycles (A) that drive daily biological rhythms (B). Nearly all cells in the body have a circadian clock, a near twenty-four-hour rhythm of activity that must be synchronized with the environment by signals from the eyes. This clock depends on a negative feedback system that regulates gene expression (C). The brain region called the suprachiasmatic nucleus of the hypothalamus (SCN) serves as the master pacemaker and sends out biochemical signals to keep the other body clocks synchronized. This figure is adapted from Takahashi (2017) with permission of the publisher, Springer Nature. © 2019 Joan M. K. Tycko.

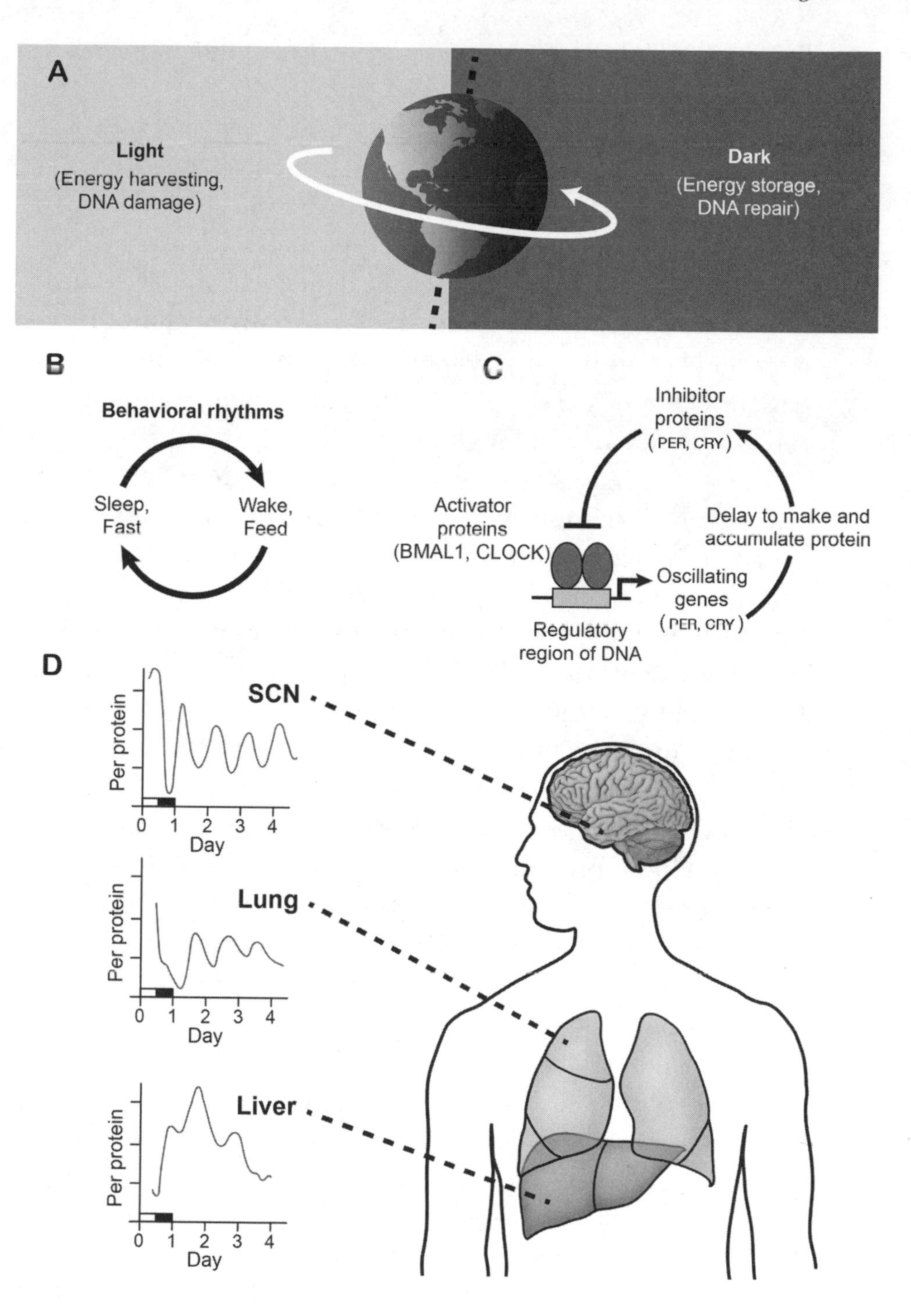
A
Light
(Energy harvesting, DNA damage)
Dark
(Energy storage, DNA repair)
B
Behavioral rhythms
Sleep, Fast
Wake, Feed
C
Inhibitor proteins
(PER, CRY)
Activator proteins
(BMAL1, CLOCK)
Delay to make and accumulate protein
Oscillating genes
(PER, CRY)
Regulatory region of DNA
D
SCN
Lung
Liver
Per protein
Day
0
1
2
3
4

the circadian clock (from the Latin *circa*, meaning approximately, and *dies*, or day).

A tiny structure within the brain called the suprachiasmatic nucleus of the hypothalamus (which means "above the place where the optic nerves cross" and is abbreviated SCN) is the body's master timekeeper. Laboratory animals like mice and monkeys that sustain damage to the SCN no longer have normal sleep-wake cycles (or any of the other circadian behavioral or physiological rhythms). Rather, they have brief periods of sleep and waking distributed randomly throughout the day and night.

For our purposes here, we don't need to delve into the fine details of the molecular machinery that produces the circadian oscillation. In simplified terms, it works like this (illustrated in figure 15C). There are a set of genes that instruct the production of proteins with names like PER and CRY. These genes are turned on by activator proteins called BMAL1 and CLOCK, which work together. The crucial link to complete a loop of signals is that PER and CRY proteins feed back to inhibit the gene activation produced by BMAL1 and CLOCK. Because it takes a while for enough PER and CRY protein to accumulate in the cell to start inhibiting their targets, the amount of these proteins oscillates up and down, and it turns out that this feedback system is tuned to cycle every 24.3 hours or so. There are many more details of the circadian clock, but that's the basic idea: it works through a negative feedback loop of gene expression.[13]

Light coordinates the timing of the internal circadian clock with the external world through light-sensing neurons in the retina. These include a group of large, spindly cells called intrinsically photosensitive ganglion cells. These neurons send their axons to the SCN to convey electrical information about the overall ambient light level. This stream of information from the eyes produces subtle daily adjustments in the SCN's master clock. The SCN neurons then communicate this information to all of the tissues in the body, using both neural signals and circulating hormones (figure 15).[14] In this way, activities in the various tissues of the body are at least approximately synchronized with the solar cycle. It's not perfect: the kidney clock runs at about 24.5 hours, whereas the cells in the cornea oscillate with a period of about 21.5 hours. This rough synchrony appears to be good enough for healthy body function.

OVER TWENTY YEARS AGO, an adult patient entered a sleep clinic in Salt Lake City with a disabling problem. She was such a lark that she began to feel sleepy in the early evening. She was typically asleep in bed by 7:30 p.m., only to awaken at 4 a.m. She wasn't able to have much of a social life with that schedule. During her treatment, she mentioned that several other members of her extended family had the same problem. There's no statement that's more likely to get the attention of a geneticist than that, and so Louis Ptáček and his colleagues soon tracked down her relatives. Eventually, they found twenty-nine people in three families with this extreme lark behavior, which they named familial advanced sleep phase syndrome (FASPS). This rare trait is dominant, so you only need to inherit a copy from one of your parents to have it. When FASPS patients were analyzed, it was found that many of their body rhythms were shifted three to four hours earlier, including the nighttime low point in body temperature and the time of day when the hormone melatonin begins to be secreted.[15]

A few years later, Ptáček's group joined forces with Ying-Hui Fu's lab,[16] and together they found that his FASPS patients carried a single mutation in the *PER2* gene, disrupting its function in the circadian clock.[17] Since that time, additional FASPS families have been found bearing mutations in other genes encoding parts of the circadian clock: *CRY2*, *PER3*, and *CK1DELTA* (which interacts with the PER proteins). As another test, when these various mutations in circadian clock genes were introduced into mice using genetic engineering, the mice showed early-shifted activity and body temperature rhythms, similar to the FASPS patients. These results are very satisfying, but we should not rush to conclude that all extreme larks result from mutations in core circadian clock genes. There are some people with this trait who do not appear to have mutations in any of those genes.

Another rare sleep trait that runs in families is short sleep duration. The patients who have this trait are called familial natural short sleepers. They sleep about six hours per night and have no detrimental health consequences. Two different short-sleeping families have been analyzed, both of which bear mutations in the *DEC2* gene, which may or may not be a regulator of the circadian clock system (there are conflicting reports in the literature) and for which the relationship to the brain's sleep circuitry remains unclear.[18] Another short-sleeping family was found to harbor a mutation in the *ADRB1* gene. When Fu and Ptáček genetically engineered mice to bear this mutation, they too became short sleepers. Here, the link between the gene and the brain's sleep circuits is better understood. *ADBR1* directs the production of a neurotransmitter receptor called the beta-1-adrenergic receptor, which is involved in regulating the electrical activity in a brain stem region called the pons that is important in the transition from sleep to wakefulness.[19] On the other end of the sleep duration distribution, when a particular mutation is created in *SIK3*, which is important for regulating the clock gene *PER2*, extreme long-sleeping mice are produced.[20] However, as of this writing, it is not yet clear if genetic variation in *SIK3* affects the duration of human sleep.

Both FASPS and familial natural short sleep are rare conditions that fall into the left-hand tails of the bell-shaped distributions for chronotype and sleep duration shown in figure 14. Could gene variation also underlie subtler variations in sleep found in the central portions of these two distributions? There are two lines of evidence that suggest a genetic component to normal sleep-cycle variation. First, a pilot study that used Fitbit activity trackers to monitor the sleep of identical versus fraternal twin pairs estimated that about 50 percent of the variation in sleep duration and about 90 percent of the variation in restless sleep incidence was heritable.[21]

Second, three recent large GWAS investigations sought to find gene variants associated with the continuum of lark-to-owl chronotypes.[22] Impressively, variation in four genes was implicated in all three studies. Of these four, the aforementioned *PER2* and another gene called *RGS16* are known components of the circadian clock; a third gene, *FBXL13*, might be a regulator of the circadian clock (the literature remains unresolved on this point); and the last, *AK5*, has no known relationship to the circadian clock at all. In thinking about *AK5* and other non-clock genes (which came up in two or three of the GWAS investigations), we should remember that the circadian clock is reset by light signals from the retina. As a result, variation in genes involved in that resetting process, perhaps those expressed in the intrinsically photosensitive ganglion cells that convey the luminance signal from the retina to the SCN, might also give rise to differences in chronotype.

Eugene Aserinsky was a graduate student in the sleep laboratory of Nathaniel Kleitman at University of Chicago, where he made EEG recordings from adults as they fell asleep. These recordings showed that after falling asleep, the EEG gradually changed from a noisy, jittery trace to one with large, slow oscillations. At this point, the researchers assumed that deep sleep had been achieved and it would remain in force until waking. Their standard procedure was to record for forty-five minutes to capture this transition to slow-wave sleep and then turn the EEG recorder off to save the chart paper that was piling up on the floor in huge drifts. One night in 1952, Aserinsky had the clever idea to bring his eight-year-old son, Armond, into the lab to be that night's subject. About thirty minutes after Armond had fallen asleep, his father was watching the pens on the EEG chart recorder register the large, slow oscillations of deep sleep. Then, to his enormous surprise, the EEG shifted to a rhythm that looked more like the waking state, even though Armond was still clearly sleeping and totally immobile. This stage of sleep, associated with rapid eye movement (REM), does not usually occur in adults until about ninety minutes after falling asleep. However, in children like Armond, it occurs much sooner.

The publication of these findings by Aserinsky and Kleitman in 1953 was a watershed moment for sleep research, and in the following years a much more detailed picture of sleep emerged. When scientists left their EEG machines on all night (piling up gargantuan stacks of chart paper in the process), they found an adult sleep cycle of about ninety minutes. This consisted of the aforementioned gradual descent into deeper and deeper sleep, accompanied by gradual synchronization of the EEG. These stages of sleep are collectively called non-REM sleep, and they are further subdivided into four stages, ranging from drowsy or nodding off (stage I) to deep sleep (stage IV). Stage IV is followed by the transition into a period of REM sleep. The termination of the

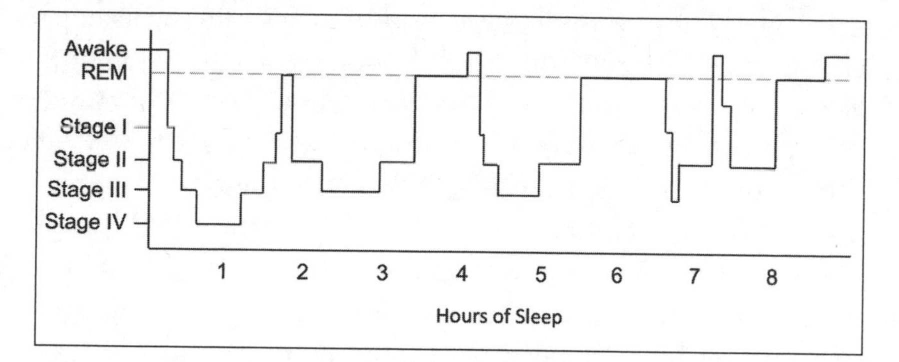

FIGURE 16. Stages of sleep throughout the night in an adult. Note that REM sleep predominates in the second half of the night. This example shows four sleep cycles.

REM sleep period marks the end of one sleep cycle (figure 16). A typical good night's sleep will consist of four or five of these cycles. As the night progresses, the composition of each sleep cycle changes so that there is proportionally more REM sleep. In the last cycle before waking, as much as 50 percent of the sleep cycle may be devoted to REM. Humans show changes in sleep over the lifespan, with the proportion of the time spent in REM sleep decreasing from about 50 percent at birth to 15 percent among the elderly.

There are a host of physiological changes that accompany REM sleep, including increases in breathing rate, heart rate, and blood pressure, and a sexual response—penile erection in men, erection of the nipples and clitoris together with vaginal lubrication in women. Even more striking are changes in muscle tone. In the course of a night, the typical adult sleeper will change his or her position about forty times, mostly without being conscious of this action. However, in REM sleep, there is no movement at all because the body goes totally limp. As a result, is almost impossible to have REM sleep in anything other than a horizontal position.

REM sleep is sometimes called "paradoxical sleep" because the EEG resembles the waking state, yet the sleeper is essentially paralyzed. The movement-commanding centers of the brain are actively sending signals to the muscles, but these impulses are blocked at the level of the brain stem by an inhibitory synaptic drive from another part of the brain and so never reach the muscles. This blockade only affects the outflow of motion commands down the spinal cord, not those of the cranial nerves, which exit the brain stem directly to control eye and facial movements (and heart rate, through the vagus nerve).[23] A failure in this blockade is seen in a condition called REM sleep behavior disorder, which is characterized by violent, presumably dream-enacting behaviors during REM sleep, often causing self-injury or injury to others. This may include punching, kicking, leaping, or even running from the bed. REM sleep behavior disorder is different than conventional sleepwalking, which only occurs during non-REM sleep.

NARCOLEPSY IS A RARE disorder that usually starts in the teenage years. Over the course of a few days or weeks, the urge to sleep during the day gets stronger and stronger. Despite having a good night's sleep and normal alertness soon after waking, these kids fall asleep in class, while doing their homework, and, dangerously, while driving. Sometimes this onset of daytime sleepiness is associated with weight gain. In narcolepsy, REM sleep can occur suddenly, at any time, without first passing through the stages of non-REM sleep. It is accompanied by dreamlike hallucinations.

If the symptoms stop there, the diagnosis is type 2 narcolepsy.[24] However, in a different form of the condition, type 1 narcolepsy, an additional terrifying symptom then develops. Here's an excerpt from a typical case report:

> An 18-year-old man presented with a 2-year history of severe excessive daytime sleepiness that began a few months after a vaccination for influenza A (H1N1). He had to repeat a school year because of the excessive daytime sleepiness. He was overweight and had frequent sleep paralysis and nightmares. Six months after the onset of excessive daytime sleepiness, he developed episodes of weakness in his limbs and neck triggered by various emotional stimuli, especially when he laughed with his brother. His brother had made a cell phone video that the neurologist was able to view, which led to the diagnosis of typical cataplectic attacks.[25]

Cataplexy, as found in type 1 narcolepsy, is a form of flaccid paralysis, typically lasting for a few seconds to two minutes, and usually begins in the face and neck and sometimes spreads outward to include the trunk and limbs.

Type 1 narcolepsy is associated with a complete loss of a small population of neurons in the lateral hypothalamus of the brain. These neurons produce the neurotransmitter orexin, which is important in regulating both alertness and feeding behavior. This explains while many patients gain weight as their narcolepsy develops. Type 2 narcolepsy involves only a partial loss of these neurons. In both types, the orexin-using neurons are destroyed by an autoimmune attack. There was an unusual increase in the incidence of type 1 narcolepsy coincident with the 2009–2010 H1N1 flu pandemic. It is suggested that some protein fragment in the H1N1 flu, or the vaccine designed to trigger immunity to it, cross-reacts with a protein on the surface of orexin-using neurons, leading to their destruction by T cells of the immune system.

Type 1 narcolepsy is a great example of gene-environment interaction. It is weakly heritable, with only about 1 to 10 percent of cases running in families. If your identical twin has narcolepsy, your chance of having it will be about 25 percent, which is much higher than the 0.05 percent overall rate in the population but far less than one would expect for an entirely heritable disease.[26] Yet, more than 98 percent of patients with type 1 narcolepsy have a particular variant of an immune-system gene (with the mind-numbing name *HLA-DQB1*0602*), compared to only 12 percent of the general population. The most likely explanation is that, in order to develop type 1 narcolepsy, you must have both the unlucky variant of the *HLA* gene and be exposed to some antigen, like the H1N1 flu virus, that triggers the autoimmune response against the orexin neurons.[27]

═══

EVEN IN SLEEP, WHEN it's largely disconnected from sensation, the human brain can generate a complete world of conscious experience, all by itself. Some mornings you might awaken with no recollection of any dreams at all, while other times the night seems to be crowded with them. In general, unless you wake up during or within a few seconds of the end of a dream, you are unlikely to recall it. Once awake, the memory of a dream will often quickly dissolve unless written down, recorded, or spoken to someone else.

Dreams are dominantly visual, in full color and movement. They are built from the prior experience of waking life, with recognizable people, places, and objects. Sounds and speech are also common in dreams, but the senses of touch (including pain and temperature), smell, and taste tend to be diminished. Most dreams have a truly sensory character—they are not mere thoughts or ruminations, and yet they are famously different from the experience of waking life.

For many years it was thought that dreaming only occurred during REM sleep. Now we know that dreams can be reported following awakening from any stage of sleep, but that their character and duration tend to vary with different sleep stages. Dreams that occur in stage I, right after sleep onset, are usually brief and, although they have a strong sensory component, this sensation does not progress to form a continuing narrative. These dreams are typically scene fragments, without much detail and with little emotional content. They tend to be logical and congruent with waking experience. Importantly, sleep-onset dreams are very likely to incorporate experiences from the previous day's events. In a classic study, subjects played the video game *Tetris* for several hours before sleep. When fitted with EEG recording gear and then awakened during stage I, more than 90 percent of the subjects reported scenes from the game, but only when they were awakened shortly after sleep onset, not during deep non-REM (stages III and IV) or REM sleep.[28]

The dreams that we tend to remember are the narrative ones that unfold in a storylike fashion and are rich in detail. They often incorporate weird blending and transformations of familiar people and places, and violations of physics like human flying. In our dreams, we can experience a different reality, but we have little control over it. We cannot pursue goals; instead, we are carried along and tend to accept sudden transformations and impossible objects as a given. There is a suspension of disbelief about otherwise illogical or bizarre experiences. Narrative dreams can often have strong emotional content, both positive and negative, but they don't have to.

Narrative dreams are the ones we are most likely to remember and discuss. In part, this is because they make for good stories (though sometimes not as interesting to others as to the dreamer), but it also reflects the structure of the sleep cycle: you are most likely to awaken, and therefore remember your dream, toward the end of the night, when

REM sleep predominates. While narrative dreams are most frequent during REM sleep, they are not exclusive to it. In sleep labs, narrative dreams can sometimes be recalled from awakenings during sleep stages III and IV, particularly during the second half of the night. Sometimes narrative dreams can be recalled when subjects are roused from naps in which no REM sleep has occurred. People with neurological damage to the brain stem, who have lost the ability to enter the REM stage, can still have narrative dreams.[29] Conversely, people in REM sleep aren't necessarily dreaming. On average, 20 percent of awakenings from REM sleep result in no dream report at all in adults, even in the lab where sleepers are awakened and immediately queried about their dreams.[30]

During REM sleep, information from the senses is almost, but not entirely ignored, due to a partial blockade of incoming electrical signals at the level of the thalamus. Olfactory information, because it does not pass through the thalamus, is still processed during REM sleep and so can serve as the basis for subconscious associative learning conducted entirely during sleep.

Nonolfactory stimuli, like the buzz of an alarm clock, can sometimes be incorporated into the end of a REM-stage dream. Rarely, stimuli like a spray of water on the face or pressure on the limbs can find their way into dream content.[31] In one provocative set of experiments that would probably not be approved today, subjects fell asleep with their eyelids taped open, and objects were illuminated in front of their eyes during the REM stage. Even these manipulations did not reliably intrude on dream content.[32]

There are several lines of evidence to suggest that narrative dreaming is more related to imagination than perception. Dreaming is abolished in some patients who sustain damage to a part of the brain called the tempero-parietal-occipital (TPO) junction. These patients also have serious deficits in mental imagery while awake. In adults, the cognitive skill that most predicts dream recall is visuospatial imagery.[33] Brain-imaging studies performed during REM sleep, when narrative dreams are most likely to occur, show activation in brain regions involved in mental imagery and visual or auditory memory. Interestingly, the brain regions involved in the initial processing of sensations like vision, touch, and sound are largely inactive during REM sleep. Also inactivated in REM sleep are regions associated with voluntary control (right inferior parietal cortex), self-monitoring, and reflective thought (posterior cingulate cortex, orbitofrontal cortex, and dorsolateral prefrontal cortex). Overall, the picture from both brain lesions and brain imaging shows that narrative dreaming is most similar to random imaginative thought in the waking state—that is, daydreaming.

Some people can report their narrative dreams every morning and others claim to rarely dream. However, if people who report rarely dreaming are taken to the sleep lab, awakened during REM sleep, and queried, most of them will report narrative dreams. As I mentioned earlier, some people with certain types of brain damage are truly lacking in narrative dreams, but they are few and far between. Most people who report rarely dreaming can train themselves to recall more dreams if they so choose. A good way to start is to keep a pad of paper or a voice recorder by the bed in order to make a note of dreams upon awakening. This will result in both more recall and, over time, more brief awakenings in the second half of the night to allow for potential dream recall.

Because dreams are formed from our past experiences, the raw material for our dreams is limited by our sensory worlds. For example, if you are born blind, you will not have visual memories and so you will not have visual dreams. However, if you are born sighted and become blind after age five or so, either from damage to the eyes or from damage to the early visual-processing stages of the brain, then you will be able to have visual dreams, built from the visual memories you have stored, albeit sometimes morphed and recombined in bizarre ways.

While you can still have visual dreams after damaging the early parts of the visual pathways in the brain, damage to the later parts of these processing pathways will affect dream content. For example, people who, as adults, have acquired brain lesions that impair their perception of faces when they are awake will not dream of particular faces. Likewise, people who have sustained damage in brain regions involved in color or motion perception have corresponding deficits in their dreams.[34]

The transitions between sleep stages are controlled by a complex interplay of neurotransmitters, including acetylcholine, norepinephrine, serotonin, histamine, and dopamine. We know that drugs that target these neurotransmitters can affect dreaming. For example, most antidepressants, which act on serotonin and norepinephrine signaling, seem to reduce the frequency of narrative-dream recall; some SSRI antidepressants enhance the emotional content of recalled dreams. Patients with Parkinson's disease, which attenuates dopamine signaling, report reduced emotional content and bizarreness in their dreams.[35] These findings make it likely that genetic variation in the receptors for these sleep-altering transmitters—as well the enzymes that make, degrade, and store them—could affect dream patterns, but at present, there's not much solid evidence either for or against this idea.

Although we might like to imagine that, in our dreams, we can be fundamentally different from our waking selves, this does not appear to be entirely true. While dreams can meld and stretch and recombine our past experiences and inject fantastical alterations of time and place, the raw material from which dreams are made is our waking experience. Not only that, formal content analysis of many dreams, across cultures, has revealed that the predominant concerns, mood, curiosity, and cognitive capacity are highly correlated between our waking selves and our dreaming selves.[36] The best predictor of the content of your dreams is your waking life.

EIGHT

A Day at the Races

SCATTERED ACROSS THE OCEANS OF SOUTHEAST ASIA live several culturally and linguistically distinct groups of sea nomads who rely on breath-hold diving to obtain their food. One of these is the Bajau people, who live in Indonesia, the Philippines, and Malaysia.[1] Another is the Moken people, who inhabit the islands off the west coast of peninsular Thailand and Myanmar.[2] Sea nomads learn to swim early in life, sometimes even before they can walk. Women, men, and children all regularly dive for harvesting, with the men doing most of the spearfishing and the women and children gathering most of the mollusks and edible worms. Most often, sea harvesting occurs at depths of fifteen to twenty-five feet, although deeper dives are also undertaken. The average adult sea nomad spends up to five hours per day submerged, the longest of any known breath-hold divers. They do all of this without the aid of wet suits, weights, or breathings aids. While, at present, the Bajau use goggles, the Moken, particularly children, often dive without them (a generation ago, none of the sea nomads used goggles). The Moken and Bajau were practicing breath-hold diving at the time of first contact with European colonizers in the sixteenth century and almost certainly had been doing so for many hundreds of years previously.[3]

All humans share a set of physiological diving responses with aquatic mammals like sea otters and dolphins. Upon entering the water, heart rate is slowed to reduce oxygen consumption and extend dive time. In addition, blood flow is routed away from organs that can withstand temporary loss of oxygen and toward the organs that need it most: the heart, brain, and active muscles. Finally, the spleen contracts to release about a half cup of red blood cells into circulation, thereby boosting the total oxygen-carrying capacity of the blood. All of these responses work together to extend the period of human breath-hold diving.[4]

To determine if the sea nomads had become genetically adapted to their particular aquatic lifestyle, Melissa Ilardo and her coworkers from the University of Copenhagen collected DNA samples from Bajau people in the village of Jaya Bakti on the island of Sulawesi. For comparison, they also collected DNA from Saluan people, who live in Koyoan, about fifteen miles away on the same island. The Saluan have little to do with the marine environment and do not engage in breath-hold diving. The researchers also brought along a portable ultrasound machine and used it to measure spleen size in all of their subjects. They found that, on average, Bajau people have larger spleens than Saluan people, and that the increase in spleen size was predicted by variation in a gene called *PDE10A*. Spleen enlargement provides the Bajau with a larger reserve of red blood cells to inject into the bloodstream during breath-hold diving. In addition, the Bajau tend to carry variants of the gene *BDKRB2* that enhance the diving-evoked reduction of heart rate and the diversion of blood flow from the periphery to the core of the body.[5]

As we all know from opening our eyes underwater, the view in this situation is blurry. Human vision is poorly adapted to the aquatic environment. However, when Moken children were tested for underwater vision, their ability to discern visual detail was about twice as good as age-matched European children—that is, their underwater vision was still blurry but not as blurry as the European kids. There are two ways to produce this partial improvement in underwater vision. One is to constrict the pupil more fully to create greater depth of field, and the other is to warp the lens to better focus the incoming light, a process called ocular accommodation. The eyes of Moken children employ both of these strategies to improve underwater vision.[6]

To date, no genetic variants have emerged that would explain the increased accommodation or pupillary constriction in Moken people or other sea nomads. In fact, it's likely that the improved underwater vision in Moken children can be entirely accounted for by practice. After eleven training sessions in a pool, spread over one month, European kids attained the same improved underwater visual acuity as Moken kids. This improvement was still evident when they were tested eight months later.[7] The lesson from the sea nomads is clear. Lifestyle and environment can produce evolutionary changes in the genes of local human populations subjected to particular adaptive pressures, as sea harvesting did with selection for spleen size and diving response among the Bajau. But, importantly, as revealed by the case of enhanced underwater vision, just because we see a significant difference in a useful trait in a particular population, that does not mean that there is necessarily a heritable basis for that difference.

═══

Modern humans arose in Africa about three hundred thousand years ago, and over the past eighty thousand years or so have spread across the world to occupy nearly every type of natural environment. Within the last twelve thousand years, most, but not all, human populations have moved from a hunter-gatherer lifestyle to one dependent on domesticating and raising plants and animals. The challenges of particular human lifestyles and locations have resulted in local selection pressures for certain traits,[8] like the enhanced physiological diving responses (but not the improved underwater vision) of the sea nomads.

One of the most notable recent human adaptations accompanied the domestication of cattle, which seems to have occurred separately in Africa and the Middle East about ten thousand years ago. Cattle domestication led to the strong selective pressure to drink and metabolize cow milk in adults. In most humans, the enzyme lactase, which breaks down the milk sugar lactose, decreases after weaning. However, in certain populations that have adopted extensive dairy farming, lactase expression persists into adulthood. Variations in the gene that encodes lactase, conferring persistent lactase expression, arose separately in the Middle East about nine thousand years ago and in Africa about five thousand years ago. Analysis of the DNA from ancient bones indicates that these gene variants spread from the Middle East into Europe only within the last four thousand years.[9] This example illustrates some important facts about human adaptation. First, new human traits can emerge rather quickly in evolutionary time: in this case, within a few thousand years. Second, sometimes, two separate human populations undergoing the same selective pressure can have the same type of genetic variant arise independently—in this case, variants influencing the expression of lactase in adulthood. This process is an example of what biologists call "convergent evolution."

However, in many cases, different populations exposed to the same environmental pressure will see different genetic variants arise. One good example of this process involves groups who live at high altitudes, above eight thousand feet, as in the Semien Mountains of Ethiopia or the high Tibetan plateau. In these environments, it becomes a challenge for the body to supply sufficient oxygen to crucial organs. Yet human populations in each of these high places have thrived in such difficult conditions. Tibetans have accumulated high-altitude-protective variants of genes *EGLN1* and *EPAS1*, while some Semien Mountain dwellers have accumulated variants of different altitude-protective genes, *VAV3*, *ARNT2*, and *THRB*. The names of these genes aren't important. The point is that the affected genes are different. The challenge of living in a high-altitude, low-oxygen environment can be met with different solutions—different combinations of useful gene variants in different populations.[10]

High altitude, breath-hold diving, and dairy consumption are just three of a growing list of local pressures for which human adaptive gene variants have been identified in local, well-defined populations. Some of the others include tolerance for dietary arsenic in people living in a particular region of Argentina, for malaria in many people of Central Africa, the Mediterranean, and India, for a marine-based diet in the native people of Greenland and Canada, and for cold temperature in native Siberians. It's a short list, but one that's growing longer all the time as more population-genetics research is conducted.

The local human adaptations I've mentioned so far involve variants in one or a small number of genes. But can local human adaptions also impact polygenic traits like height, which is about 85 percent heritable in Europeans and to which hundreds of genes contribute?[11] Here, the answer remains unclear. On average, northern Europeans are taller than southern Europeans. When 139 known height-influencing genes were compared in groups of northern and southern Europeans, it was found that frequencies of the height-promoting variants of these genes were consistently higher among the northerners.[12] A related study of Britons showed that many weakly acting height-promoting gene variants have been under strong selective pressure in the last two thousand years. Similar signatures of recent selection were found for increased newborn head circumference, female hip size (presumably to accommodate the larger infant heads during childbirth), fasting insulin levels, and several other polygenic traits.[13] However, in 2019, two different groups published papers calling into question the strength of the result supporting polygenic adaptation for height in Europe. With a larger and less stratified population sample, the effect was greatly reduced.[14] At present, it seems that there is polygenic adaptation for height among Europeans, but that it is a weak effect. This is an active area of investigation and this result may change further as still better sampling and statistical methods are brought to bear.

You'll note that, at present, there are no behavioral or cognitive traits on the list of local, recent human adaptations. Nearly all behavioral and cognitive traits in adults have a heritable component—usually in the ballpark of 50 percent—that almost always involves the small contribution of many genes.[15] Polygenic behavioral and cognitive traits could, in theory, be subject to local selective pressure in various populations, leading to partially heritable average population differences. But, as of this writing, there is no good evidence that this has been the case.

These examples of local, recent human genetic adaptation have now been mostly subjected to at least one round of replication and challenge from other investigators and, on the whole, they have survived quite well. Specific claims may be modified or invalidated by work to come, but the overall trend is undeniable: there are significant average genetic differences across local populations in a range of physical traits. The adaptive traits can involve changes in either a small number or a large number of genes and, at least in some cases, can emerge over a few thousand years.

Describing these results, which now fall squarely within the mainstream of thought in population genetics, is enough to get one branded as a racist in many quarters. To some well-meaning critics, such endeavors are to be condemned because they have been used by others to support racist genetic arguments. In their view, even if scientists who conduct such work have good intentions (say, to more effectively guide medical treatment decisions), reporting genetic differences underlying traits across local populations is situated on a slippery slope that is so laden with historical malfeasance that the project is better abandoned. To them, words like "population" and "ancestry" are just a politically correct construction to allow one to talk about race. For example, Angela Saini, commenting on the recent Human Genome Diversity Project, wrote, "The word 'race' had been prudently replaced by 'population' and 'racial difference' by 'human variation' but didn't it look suspiciously like the same old creature?"[16]

I am entirely sympathetic to these concerns. There's no question that the science of human variation has been and continues to be misused to rationalize racial bigotry around the world. Perversions of science, particularly population genetics, have been used to justify, either contemporaneously or retroactively, the Atlantic slave trade, the Jewish Holocaust, the genocide of Tutsi people in Rwanda, the stealing of land by colonial powers, and the systematic legal oppression of African Americans after emancipation. Sadly, this list goes on and on. Over the years, some of the leading scientists in the study of human differences—from Francis Galton, popularizer of the expression "nature versus nurture," to James Watson, codiscoverer of the structure of DNA—have actively promoted racist pseudoscientific ideas. Today's racists, from white supremacists in the United States to certain Hindu nationalists in India, all rely, at least in part, on arguments ostensibly based on population genetics to support their bigotry and policies of racial oppression.

So why not just be done with politically fraught investigations of genetic differences between populations and stick to the study of individual differences instead? In my view, population genetics is a valid area of inquiry that legitimately helps us understand human evolution and can more effectively guide medical care. But perhaps the most important reason not to throw the baby out with the bathwater is that the full weight of scientific investigation, including population genetics, is and will continue to be required to refute racist pseudoscientific arguments. These arguments are not going away, and they must be met with data, not mere assertion and certainly not by declaring the whole area of investigation out of bounds. The emerging results from population genetics are, and will continue to be, required to refute its own racist history.

WHAT ARE THE PRINCIPLES of so-called scientific racism? A version popular with white supremacists boils down to this:

1. There are broad, continent-based categories—like Europeans, Asians, and Africans—that reflect the clear division of humanity into a small number of biologically distinct racial groups. These racial groups have

remained fixed and unblended for tens of thousands of years, allowing genetic differences to accrue.

2. The varying environments corresponding to these broad racial categories have imposed different selective pressures. In one popular racist telling, the African environment has selected for high sex drive, high violence, and low intelligence, and the people of Asia have been selected for low libido and high intelligence but low morals. The peoples of Europe, particularly northern Europe, have been selected to be just right: a happy medium, thereby justifying their role as colonizers and their resistance to immigration.

3. As a result of points 1 and 2, one can predict average human behavioral and cognitive traits based on these broad racial categories. Racial traits are heritable and immutable, and so will be deeply resistant to any social intervention, thereby excusing ongoing oppression and denial of educational and economic opportunity to broadly defined "racial" groups.

Of course, pseudoscientific racists never make arguments to denigrate and justify the ongoing oppression of their own self-defined racial group. Funny how that works out. In some cases, such arguments have been used to explain a positive trait in another group, but always with a negative caveat. For example, a common pseudoscientific white-supremacist argument holds that Jews, as a result of being denied land ownership for many generations, gravitate to more quantitative, urban occupations like moneylending and shopkeeping. In this story, the ban on Jewish land ownership created selective pressure for evolved intelligence, but also instilled a devious, amoral character. Similar fake evolutionary stories are told about East Asians and work habits, or Africans and athletic prowess.[17]

LET'S EXAMINE THE CORE claims of pseudoscientific racism to see how they hold up. First up is this contention:

> There are broad, continent-based categories—like Europeans, Asians, and Africans—that reflect the clear division of humanity into a small number of biologically distinct racial groups.

If you live in the United States, you are required to fill out a form every ten years at the time of the national census. On this form, you are asked to tick one or more boxes to identify your race. Today, the choices are:

> **Black or African American**: A person having origins in any of the Black racial groups of Africa.
>
> **American Indian or Alaska Native**: A person having origins in any of the original peoples of North and South America (including Central America) and who maintains tribal affiliation or community attachment.
>
> **White**: A person having origins in any of the original peoples of Europe, the Middle East, or North Africa.
>
> **Asian**: A person having origins in any of the original peoples of the Far East, Southeast Asia, or the Indian subcontinent including, for example, Cambodia, China, India, Japan, Korea, Malaysia, Pakistan, the Philippine Islands, Thailand, and Vietnam.

Native Hawaiian or Other Pacific Islander: A person having origins in any of the original peoples of Hawaii, Guam, Samoa, or other Pacific Islands.

The instructions go on to clarify that "people who identify their origin as Hispanic, Latino, or Spanish may be of any race."[18] If you are filling out a similar form in the United Kingdom, the categories are different. The tick boxes used in 1991 were White, Black-Caribbean, Black-African, Black-Other, Indian, Pakistani, Bangladeshi, Chinese, and Any Other Ethnic Group. In the Brazilian census, you must pick a single racial category from a list consisting of *Branca* (white), *Parda* (brown, meaning multiracial), *Preta* (black), *Amarela* (yellow, Asian) or *Indígena* (indigenous). These lists of terms don't capture cultural differences in how the categories are used. For example, in the United States, people who would self-categorize as black have, on average, 80 percent West African and 20 percent European ancestry (of course, these are just averages). In Brazil, most people who identify as *preta* have nearly entirely West African ancestry. Most people who self-identify as black in the United States would call themselves *parda* if they lived in Brazil. Furthermore, people in Brazil with nearly entirely West African ancestry are more likely to identify as *preta* if they are poor but *parda* if they are more affluent.

The key point here is that racial categories are socially and culturally constructed. They are vague, changeable, and vary from place to place and culture to culture, reflecting local history and politics, in particular the legacy of colonialism. It's not a coincidence, for example, that the UK census distinguishes people with ancestry from India, Pakistan, and Bangladesh, reflecting British colonial history in that part of the world, while the US and Brazilian censuses lump them all in the general category of Asian. Some of the categories are based on present-day nations, others on continents. If we move away from government census forms and into the street, people around the world readily make racial classifications based on still more criteria, like language (Hispanic) or religion (Muslim, Jewish, Hindu, etc.) And, importantly, these categories can vary, even within the same country or community. To the US government and most people on the street, I'm white. But to the white supremacists on the internet, I'm decidedly not, due to my disqualifying Ashkenazi Jewish ancestry. Around the world, in every place or culture, racial classifications exist, and they are inherently ideological, economic, and political. Any scientific study involving race must accept and engage with this fact.

It's not that racial categories don't exist, as some well-meaning, idealistic people would claim. Rather, it's that racial categories are not biological. They are not taxonomic subdivisions of the human species. And they are not, as some would contend, analogous to purposefully created and maintained breeds of domesticated cat or dog,[19] or, for that matter, like Trut and Belyaev's tame foxes. Rather, racial categories are cultural-biological categories. They are the collision of biological human differences and the local and changing cultural decisions about what kinds of differences are important.

The fact that race is not a biological phenomenon does not make it unworthy of study. After all, education, money, social class, and reputation are all unnatural phenomena, yet we understand their importance to human lives. Anthropologist Jonathan Marks nicely unpacks the implications of ignoring race when he writes,

> We ought to be leery, then, of the statement "race doesn't exist" simply because race doesn't exist as a unit of nature, or biology, or genetics. For if the only reality we acknowledge is

> nature, what do we make of political or social or economic inequality? Those are real facts of history and society, rather than facts of nature. Do they suddenly vanish, then? If we synonymize the non-natural with the unreal, then poverty becomes not a problem to be solved, but a phantom to be ignored.[20]

IF THERE REALLY ARE genetic differences that correspond to broad, culturally defined racial groups and underlie average differences in traits, then we should be able to see them reflected in the analysis of many individual genomes. Richard Lewontin sought to address this question in 1972, years before the analysis of human DNA was possible.[21] He examined variation in blood proteins in many people around the world, whom he grouped into seven invented "races": Africans, West Eurasians, East Asians, South Asians, indigenous Australians, Native Americans, and Oceanians. He found that about 85 percent of the variation in protein types could be ascribed to variation within his seven races, and only 15 percent by variation between them. His conclusion in this paper is often quoted: "Races and populations are remarkably similar to each other, with the largest part by far of human variation being accounted for by the differences between individuals. . . . Since [such] racial classification is now seen to be of virtually no genetic or taxonomic significance either, no justification can be offered for its continuance."[22]

When Lewontin's finding was published, many argued that this was proof that racial categories were biologically meaningless. Yet this was cognitively dissonant for most people. If race is biologically meaningless, then how can we look at someone's headshot photo (without information about their voice or clothing or mannerisms) and assign them to a broad category of ancestry with reasonable, albeit not total, accuracy? Clearly, we can make these guesses based on a few external traits like skin color, facial features, eye color, and hair color and texture. Overall, most variation in traits is within races rather than between races, but clearly there are a handful of external traits, the combination of which can help inform our broad estimates of an individual's ancestry.

In Lewontin's day, it was not possible to scan many locations in the human genome for variation. Thirty years later, in 2002, Marcus Feldman and his colleagues revisited this question. They analyzed 377 variable locations in the genomes of 1,056 individuals from around the world and replicated Lewontin's basic finding: for the vast majority of individual locations in the genome, it is impossible to identify genetic variants to distinguish one broad race from another, because there is, on average, much more genetic variation within these racial groups than between them. However, when variation at all 377 locations in the genome was analyzed together with multivariate Bayesian statistics, and a computer program was instructed to group the data into five clusters, the groups that emerged were roughly consistent with some popular broad categories of race in the United States: African, East Asian, European, Oceanian, and Native American.[23]

The publication of this study and a few others that followed generated a lot of controversy. Some said that it reinforced the notion that broad racial categories are biologically true and enduring taxonomic divisions of the human species. This is incorrect. First, there's nothing magic about those five populations. The number five was generated by the experimenters, not the data. In fact, it has been argued that, with a better sampling of African DNA, a division into fourteen populations is more accurate.[24] What these types of studies show is actually rather intuitive: people are more likely to mate with those who are nearby as opposed to those who are far away, and the probability of mating is gradually reduced with distance. In addition, it shows that large geographic barriers like oceans or the Sahara desert have been impediments to mating. Where those barriers are porous, more interbreeding occurs and the population clusters, as defined by genetic markers, blur at the margins.[25] Racial categories are not biological categories, but neither are they completely arbitrary cultural constructions. They are dynamic, culturally created categories that are based on a very small subset of observable heritable physical traits. Historically, they have been used to denigrate and control millions of people unjustly.

ANOTHER KEY TENET OF pseudoscientific racism is:

> Racial groups have remained fixed and unblended for tens of thousands of years.

This assertion was a pillar of Nazi ideology, which claimed unmixed descent from the Corded Ware culture, named for the style of their five-thousand-year-old pottery retrieved from archeological sites. In the Nazi mythology, the Corded Ware people were early Aryans with deep roots in Germany. The Nazis further argued that because Corded Ware pottery was found in archeological sites in Poland, western Russia, and Czechoslovakia, that the Germans had an ancient claim on those lands. A similar myth underlies certain strains of Hindu nationalist ideology, which holds that, over many thousands of years, there has been no substantial contribution to Indian bloodlines or culture from people outside of South Asia.

The newest findings, which analyze DNA from ancient human remains and compare them to modern populations, provide a definitive rebuke to these claims of racial purity.[26] In the case of the Corded Ware culture, the ancient DNA clearly shows that they mostly derive from a mass migration of Yamnaya people from the steppes of present-day Russia and Central Asia about five thousand years ago. Suck on that, Nazis.

Likewise, the ancestry of modern-day Indians, even in south India, was formed in large part by several waves of mass migration from present-day Iran and the Eurasian steppe. In fact, about 50 percent of the genetic variation found in today's Indians derives from these migrations, which started about five thousand years ago. All over the world, nearly every population alive today is the product of repeated mixing over many thousands of years. As geneticist David Reich writes, "The story that is emerging differs from the one we learned as children, or from popular culture. It is full of surprises: massive mixtures of differentiated populations; sweeping population replacements and expansions; and population divisions in prehistoric times that did not fall along the same lines as population differences that exist today."[27] The analysis of ancient DNA shows definitively that there is no mythical racial purity. People in the United States who self-identify as white on the census are a mixture of at least four ancient populations that thrived about ten thousand years ago. Those ancient populations were as different from one another as East Asians and Europeans are now. In other words, we're all mutts.

But we're not exactly the same type of mutt. These days, you can spit in a tube, mail it off with $99, and get back a report saying that your ancestry is, for example, 33 percent Welsh, 42 percent Turkish, 20 percent Swedish, and 5 percent Greek. Or you might learn that you descend from 85 percent West African, 10 percent English, and 5 percent French ancestors. What's important to realize about these estimates of your ancestry—and they are only rough estimates—is that the categories are chosen to be culturally meaningful to present-day people.[28] They look back to an arbitrary time about five hundred years ago, before European colonialism got fully underway, thereby precipitating some of the more recent waves of genetic mixing. The direct-to-consumer ancestry companies could generate a genetic report using the categories of about three thousand years ago that might say that your ancestry is, for example, 45 percent Hittite and 55 percent Oxus, but that wouldn't be very appealing to most of their customers. Or they could go back two hundred thousand years and tell everyone that their ancestry is African. In fact, I'll tell you that for free. You don't even need to spit in a tube.

LET'S CONTINUE OUR EXAMINATION of the racist pseudoscientific argument.

> The varying environments corresponding to these broad racial categories have imposed different selective pressures. . . . The African environment has selected for high sex drive, high violence, and low intelligence, and the people of Asia have been selected for low libido and high intelligence but low morals. The peoples of Europe, particularly northern Europe, have been selected to be just right.

One of the hallmarks of racial categories is that they are broad: white, black, Asian, and so on. We talked about some examples of recent human adaptations to environmental conditions like high altitude, cold temperatures, and a marine-based diet. Crucially, these are *local* environmental conditions. When people speak of races, they are generally not talking about present-day local populations like the indigenous Siberians, who must contend with extreme cold, or the Bajau of Sulawesi, who have adapted to extended breath-hold diving. There is no specific "environment of Asia" to adapt to because, of course, Asia includes high mountains, deserts, boreal forests, beaches, tundra, grasslands, and tropical rainforests. The same holds true of the other continent-wide racial terms that are so popular. If you hold, as many racists do, that it has been so easy to obtain food and shelter in Africa that the selective pressure to be clever is much smaller than it is in Europe, then you must imagine that this pertains to all of the various local environments of Africa, or even just sub-Saharan Africa—from tropical forests to high mountains to desserts—and, in the same way, it must pertain to all the various local environments of Europe.

Even when you get down to racial categories based on present-day nations, the same problem exists. What are the unique selective pressures of the Chinese environment? Of course there aren't any, because China is composed of many distinct environments and lifestyles. The situation is even sillier when some linguistically based racial terms are used. What is the natural environment of the Hispanics? There isn't even a single continent of Hispanics! As a result, it's hard to imagine a scenario in which local selective pressures could drive average genetic changes in Hispanics as a group. Arguments that imagine different selective pressures on broad racial groups to produce different cognitive or behavioral traits do not withstand even the most cursory scrutiny.

There is almost no topic that is more contentious, either in terms of individuals or populations, than intelligence. But what is it? Psychologist Linda Gottfredson of the University of Delaware has provided a useful definition:[29]

> [Intelligence] . . . involves the ability to reason, plan, solve problems, think abstractly, comprehend complex ideas, learn quickly and learn from experience. It is not merely book learning, a narrow academic skill or test-taking smarts. Rather, it reflects a broader and deeper capability for comprehending our surroundings—"catching on" "making sense" of things or "figuring out" what to do.

Some psychologists break intelligence down into subdomains, like crystallized intelligence (a person's store of knowledge about the world, including both facts and procedures) and fluid intelligence (an ability to solve novel problems that relies minimally on stored knowledge). These are domains that various intelligence tests seek to measure, and that are sometimes expressed as an IQ score.

There are some who believe that intelligence tests are meaningless, but the weight of evidence is against this view. These tests are far from perfect, and some of them have aspects that are culture-bound. But around the world and across economic circumstance, they are reasonable predictors of success in school, advancement in the workforce, and even lifespan.[30] An IQ test score does not reflect everything that one would like to know about human intelligence, but in truth no measure could ever meet such a challenge. There are tests that seek to measure other aspects of cognition, like creativity (the ability to produce unusual solutions to problems or pose novel questions) or practical intelligence. These measures can provide some additional predictive power beyond the more widely used IQ tests, but not that much.

If we examine a histogram for adults in the United States, in which IQ test score is plotted against the likelihood of obtaining that score, we see a roughly bell-shaped distribution, with the average close to one hundred points (IQ tests are calibrated to have the average for large populations fall at

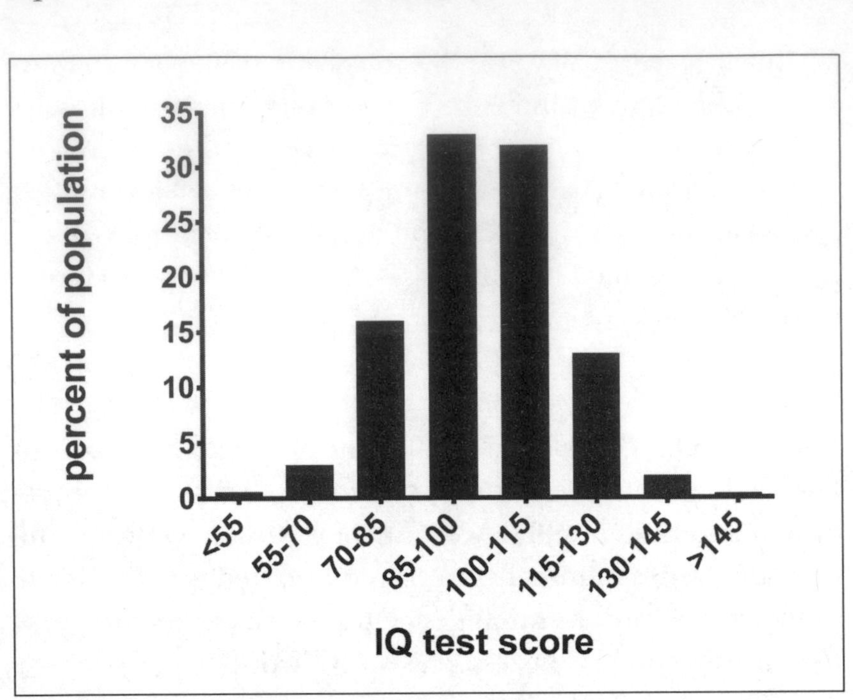

FIGURE 17. A histogram showing the approximate distribution of adult IQ test scores in the United States. The mean score is one hundred points by definition. The distribution is approximately bell-shaped (Gaussian), but with a small extra bump centered at about seventy-five points. This small bump mostly reflects highly deleterious mutations in a limited number of genes involved in brain development and synaptic function that produce intellectual disability.

that number). In this distribution, about 14 percent of the population has an IQ above 115 and about 2 percent above 130. The falloff on the lower side of the scale has roughly the same shape, but with a second, much smaller bump added, centered at about seventy-five points (figure 17). IQ scores of seventy or below are taken to indicate intellectual disability, which affects about 2 percent of the population in the United States.[31]

This distribution is very similar to what one sees for height in the United States. Height is quite heritable (about

85 percent in the MISTRA twin study we discussed in chapter 1) and also a highly polygenic trait, reflecting small variations in hundreds of genes and their interaction, both with the environment and with each other, through summation and more complex processes. However, in a small fraction of the population, there are rare single-gene mutations that can drastically effect height, like those involving the secretion of growth hormone, and can produce dwarfism or some related conditions, resulting in a small added bump at the low end of the height distribution.

The statistics for IQ test scores are very similar. The estimates for the heritable component of IQ in the United States range from about 70 percent (from the MISTRA study) to about 50 percent (using other methods).[32] At either end of this range, the conclusion is similar: IQ test scores have a significant heritable component, but a significant nonheritable component as well. Just like every other behavioral or cognitive trait, the heritable component is highly polygenic. As with height, there are a number of mutations in single genes that can have an outsize impact on IQ score. These genes include *SYNGAP1*, *SHANK3*, and *NLGN4*, and they direct the expression of proteins that are involved in the brain's wiring during early development and its ability to subtly change its electrical and synaptic properties in response to experience across the lifespan. Because mutations in these genes affect the wiring and electrical function of the brain, they often cause other neuropsychiatric problems, like autism or epilepsy, together with intellectual disability.[33]

IN THE 1960S, TESTS performed on high schoolers in the United Kingdom and the Republic of Ireland revealed that the average IQ score was about fifteen points lower in the Irish. In the years to follow, some academics, including the famous psychologist H. J. Eysenck, asserted that the gap in IQ score was due to genetic differences between the populations of these two nations. Their argument was as follows: From twin studies we know that IQ test score has a significant heritable component among both groups, so the difference between the groups must also be mostly heritable. As a result, there's no point is trying to improve the average intelligence of the Irish through better health care, nutrition, or schooling, because they will always be genetically intellectually inferior.[34]

This argument fundamentally misuses estimates of trait heritability. As we have discussed, *an estimate of the heritable component of a trait is only valid for the population in which it was studied.* If a trait has a heritable component in two different populations, that tells us nothing about the origin of the difference *between* those populations. For example, body mass index (BMI) is a highly heritable trait in both the United States and France. Average BMI is significantly greater in the United States, but the difference between the Americans and the French is not due to genetic differences between the two populations.[35] Rather, it's because, on average, Americans eat more highly caloric foods (and, to a smaller degree, they exercise less).

If the hypothesis is that the IQ score differences between the Irish and the citizens of the United Kingdom measured in the 1960s are due to genetic differences between the populations, then the main testable prediction would be that these differences should persist over at least a few generations. That prediction has been conclusively falsified. Repeated IQ testing in recent years has revealed that the Irish have improved such that there is no longer a statistically significant difference between them and the people of the United Kingdom.[36] While there is no proof that these changes are causally related, it is worthwhile to note that the average standard of living in Ireland has increased dramatically between the 1960s and the present day, with attendant improvements in health care, nutrition, and education.

A second prediction of the genetic hypothesis for population differences in IQ is that the estimate of heritability should be the same for both populations. Recall that 85 percent of the variability in height is accounted for by heritability in the United States, but only 50 percent in rural India or similarly impoverished populations because, due to malnutrition and disease, impoverished people are unable to fulfill their genetic potential for height. To my knowledge, heritability estimates for IQ test scores do not exist for Ireland and the United Kingdom circa 1960, but they have been made in many other circumstances. Several different studies in the United States have shown that the heritability of IQ test score is significantly higher in the middle class than it is in the poor, as well as significantly higher in students with well-educated parents and in people who self-identify as white compared to black.[37] These findings argue against the genetic hypothesis for IQ group differences.

Also around 1960, it was reported that in the United States the average IQ test score for self-identified blacks was about fifteen points lower than for self-identified whites. This was about the same gap as was seen for Ireland and the United Kingdom around that time. In fact, many of the same people who asserted a genetic basis for the Irish-UK IQ test gap made and continue to make the same assertion for the black-white gap in the United States. The most prominent was the best-selling book *The Bell Curve,* by Richard J. Herrnstein and Charles Murray, published in 1994. However, in opposition to this hypothesis, since 1965 the IQ test gap between blacks and whites has narrowed, falling from about fifteen points to about nine points.[38] That this gap has not been entirely erased is not surprising, as profound economic and social inequities remain between blacks and whites in the United States.

Beginning in infancy, children in wealthier families are more likely to have social experiences that contribute to their intellectual development. For example, in one study in the United States, the children of parents with professional careers were estimated to have heard thirty million words by the age of three, while the children of working-class parents heard only twenty million words, and those words were drawn from a smaller vocabulary. This study also showed that the children of working-class parents received fewer encouraging comments and more reprimands.[39] Similar studies have shown, not surprisingly, that kids from wealthier families have greater access to books, newspapers, and computers.

While it is likely that these social experiences influence children's intellectual development, it is difficult to know how much of the effect is due to the environment and how much is due to the gene variants that the parents who create these environments pass on to their children. Disentangling these effects requires adoption studies, and their evidence is clear. When children are adopted into homes of higher socioeconomic status, they achieve, on average, a twelve-to-eighteen-point boost in IQ test score, compared with siblings who were not adopted or who were adopted into homes with lower socioeconomic status.[40] Of course, the home environment is not the whole story behind this improvement, as families with higher socioeconomic status also tend to benefit from better schools, better health care, and safer, less traumatizing neighborhoods. Nevertheless, these adoption results argue against the claim that social interventions fail to close racial IQ gaps because IQ is mostly genetically determined.

In many ways, IQ is much like any other behavioral trait. IQ test score has substantial heritable and nonheritable components, and the balance between the two can vary between populations, becoming more heritable in wealthier populations with more social, economic, and political power. The nonheritable components of IQ include social experiences both within and outside the family, nonsocial experiences like nutrition and infection, and, of course, a dose of randomness from the imprecise, stochastic nature of brain development. While there are some mutations in single genes associated with brain development and plasticity that can produce large cognitive disruptions,[41] the heritable component of the variation in IQ over a score of eighty, where most people reside, is highly polygenic.

There are now GWAS investigations seeking to discover gene variants associated with general cognitive function, and they have enrolled sufficient numbers of subjects to achieve reasonable statistical power.[42] These studies have assessed general cognitive function with IQ tests or similar instruments biased toward assessment of fluid intelligence. GWAS endeavors, which have enrolled mostly residents of European countries with varying ancestry, have revealed that variation in over one thousand genes contributes to general intelligence. Not surprisingly, these genes tend to be expressed in the brain, and many are related to the development, synaptic function, and electrical activity of the nervous system. Together, variation in all of these identified genes accounts for about 30 percent of the variability in the test scores for the sampled population.

It's important to reiterate that none of these thousand-plus genes are specialized to confer intelligence. They code for proteins involved in, for example, letting ions flow across the membrane of neurons to create essential electrical signals, or guiding the growing tips of neurons to connect to their neighbors. It shouldn't be surprising that the gene variants that promote intelligence also have other effects in the nervous system. On average, they protect against Alzheimer's disease and depression. But they also put one at higher risk for autism.[43] Interestingly, intelligence-related gene variants are not just expressed in the nervous system. For example, one such gene, *SLC39A8*, directs expression of a protein that uses energy to move zinc ions (Zn^{2+}) across the membrane of cells. This zinc ion transporter protein is found in neurons but is even more strongly expressed in cells of the pancreas. So the variant of *SLC39A8* that, on average, tends to provide a tiny boost to intelligence probably also has an effect on pancreatic function. The point here is that, when we think about gene variants that affect intelligence, we have to recognize that they have many other effects, across the whole body, most of which we don't understand.

LET'S MOVE ON TO the final tenet of pseudoscientific racism:

> One can predict average human behavioral and cognitive traits based on these broad racial categories. Racial traits are heritable and immutable, and so will be deeply resistant to any social intervention, thereby excusing ongoing oppression and denial of educational and economic opportunity to broadly defined "racial" groups.

If, as is widely held by racists, genetic differences account for a large share of the remaining IQ test score gap between blacks and whites in the United States, then there must be average differences in the prevalence of gene variants, starting with the thousand or so intelligence-related genes that have been revealed by the large GWAS investigations.

Furthermore, those average racial differences must be sufficient to account for a large part of the remaining nine-point gap in IQ test score.

I can't say this loudly enough: *There is no evidence for significant average differences in intelligence-related genes between "races."*[44] *Not between self-identified whites and blacks in the United States, nor between any pair of self-defined racial groups.* Not only that, there is no evidence for racial group differences in genes that have been linked to *any behavioral or cognitive trait.* Not aggression. Not ADHD. Not extraversion. Not depression. Nada, *niente, nichts,* bupkis.

Science is not about what *could* happen; it's about what we can prove *did* happen. To assert that genetic variants underlying "racial" differences in cognitive or behavioral traits must exist because of some just-so story about continent-wide selective pressure, without providing the genetic evidence, is nonsense. It is the very definition of nonscientific, self-serving racial bigotry.

Epilogue

WHAT DOES THE SCIENCE OF HUMAN INDIVIDUALity tell us about free will and human agency? Are we genetically predetermined automatons, directed by our gene variants to have particular maladies, personalities, skills, intelligence, and sexual desires? Or are we pure blank slates, ready to be individually formed by our social and cultural experiences into shining creatures of free will, with limitless potential and choice? Of course, the answer is neither one. As we've discussed, a good phrase to replace the tired and inaccurate "nature versus nurture" is the more complicated "heredity interacting with experience, filtered through the inherent randomness of development." Experience, in this sense, is a broad category that includes social and cultural influences, but also the illnesses you've had, your physical environment, the bacteria that have colonized your body, and even potentially the cells from your mother and your older siblings (and, for some women, the fetuses you have carried) that may still live in your body.

Some human traits are highly heritable and, of those, a few derive from variation in a single gene (like earwax type) or a fairly small number of genes (like eye color), but most others are polygenic (like height) and so reflect the interaction of variation in hundreds of genes. Yet other traits have little or no heritable component at all (like political beliefs and speech accent, respectively). Most traits, whether they are behavioral (like extraversion or fluid intelligence) or structural (like BMI or propensity for heart disease) come from a mixture of heritable and nonheritable factors. Behavioral and cognitive traits are highly polygenic, so there's no single gene variant that accounts for shyness or creativity or aggression or ADHD.

Importantly, heritable and nonheritable factors can interact. This interaction can occur in a simple way: to get the disease PKU, you have to both inherit two broken copies of a gene for phenylalanine metabolism and eat foods that contain phenylalanine. Genes and environment can also interact through behavior. For example, if you're born with gene variants that make you a fast runner, then you are more likely to engage in the sports for which this is an advantage, and that practice will make you even better at your chosen sport. The central point here is that genes and experience do not always work in opposition—rather, in some cases, they can reinforce each other.

Overall, adults in the United States are pretty good at estimating the heritable components of traits. In one recent online survey, most people guessed more or less correctly that, for example, political beliefs have a very small heritable component, height is strongly heritable, and musical talent falls in the middle. There are a few traits for which people's estimates tend to be inaccurate. For example, most people think that variation in sexual orientation is about 60 percent heritable, whereas it's really only about 30 percent (about 40 percent in men and 20 percent in women). On the other side, most people think that variation in BMI is about 40 percent heritable, when it's really about 65 percent.[1] It's interesting to imagine the ways in which cultural ideas inform these mismatches. In the case of BMI, I imagine that many people want to believe that food consumption is more a matter of personal willpower than it really is. This is a frequent, if mysterious theme. In most cases—from the accuracy of memory to the heritability of personality traits—people imagine that they (and others) have a greater degree of autonomy and personal agency than they really do.

THESE DAYS, THERE'S A lot of excitement and discussion about the prospect of using gene-editing techniques (particularly one called CRISPR-Cas9) to reverse, in embryos, certain genetic diseases that are produced by variants in one or a small number of genes. In 2018, He Jiankui, of China's Southern University of Science and Technology, reportedly violated institutional approval and informed consent procedures when he deleted the *CCR5* gene in human embryos. The embryos were then implanted into their mother, resulting in the birth of twin girls named Lulu and Nana. Ostensibly, the medical justification for this procedure was to ensure that the embryos would not become infected by HIV, which their father carried. In the absence of the protein directed by the *CCR5* gene, HIV cannot gain entry into immune cells to infect them. He has been broadly condemned for this experiment. In addition to the issues of approval and consent, there is concern that *CCR5* deletion will have unintended consequences for the gene-edited twin girls. We know, for example, that *CCR5* is expressed in the brain, but its function there is poorly understood. It's entirely possible that there will be neuropsychiatric changes that result from *CCR5* deletion.

So, in addition to correcting genetic diseases or preventing infection, would it be possible to use CRISPR technology to change other traits? The answer is yes for traits specified by one or a few genes. For example, it would not be a technical challenge to ensure that your child had the variant of the *ABCC11* gene that conferred wet earwax and stinky armpits. Eye color, which is mostly controlled by two genes (but which shows minor effects from fourteen more) is another human trait that, with a bit more effort, could potentially be manipulated by gene editing. However, as with *CCR5*, even manipulating small numbers of genes may produce unintended effects. For example, the variant of *ABCC11* that confers wet earwax has been suggested to confer a slightly higher risk of breast cancer.[2]

Most of the traits that people would like to enhance in their children—like height, athleticism, and intelligence—are highly polygenic. In addition to important ethical considerations, that's a technical problem on several levels. First, it's not feasible to edit, for example, all one thousand or so presently known intelligence genes (which together only account for about 30 percent of the variation in IQ test score). Second, because these mutations do not just sum up, each adding a tiny bit to intelligence, it's not necessarily clear how to produce the best combination. A variant of gene X may be associated with increased intelligence, and another variant of gene Y might be associated with increased intelligence, but when the two are expressed together, something unpredictable could happen. The double variant could reduce intelligence, or increase intelligence but produce epilepsy, or even create some medical problem that's unrelated to the nervous system. Multiply that issue by a thousand and you see the scale of the problem. Our present state of genetic knowledge is such that it's much easier to break a sought-after polygenic trait than to enhance it. At present, we know of a few single-gene mutations that will confer intellectual disability, but none that will strongly enhance intelligence.

When you think about the individual genetic and developmental differences that impact the sensory portions of our nervous systems, it's remarkable that we can agree on a shared reality at all. You'll recall that 30 percent of the four hundred or so olfactory receptor genes are functionally different when comparing two random individuals. That's the first step of the sense of smell, before we even consider individual differences in the brain circuits that process that information or the ways in which those brain circuits are changed by experience. Due to these innate and learned differences in smell and taste perception, my integrated flavor experience of Barolo wine or Cheez Whiz is not exactly the same as yours.

Crucially, these individual differences in perception are present for all sensory systems, not just smell and taste. My red is not necessarily your red, my G-minor chord is not your G-minor chord, and my chilly bedroom is not your chilly bedroom. This individual variation doesn't hold just for those senses that point outward, but also for those that point inward and inform us about the state of our bodies. In this spirit, my sensation of a full stomach is not your sensation of a full stomach and my ten-degree leftward tilt of the head is not your ten-degree leftward tilt of the head. Each of us operates from a different perception of the world and a different perception of ourselves.

A portion of the individual variation in sensory systems is innate. But those innate effects are elaborated and magnified with time as we accumulate experiences, expectations, and memories, filtered through and in turn modifying those very same sensory systems. In this way, the interacting forces of heredity, experience, plasticity, and development resonate to make us unique.

Acknowledgments

When I'm bursting to share a cool new factoid I've unearthed for this book, the people in my academic department can see it on my face. It speaks well of them that, despite being accosted in the elevator on a regular basis, they've managed to keep their eye-rolling to a minimum.

"Did you know that nine-banded armadillos are born as identical quadruplets?"

"I just learned that 0.25 percent of twin pairs born to married women in the United States have different fathers!"

They are saints, really. For their forbearance, curiosity, and insightful questions throughout this process, I thank the entire community of scientists at Johns Hopkins University School of Medicine, and, in particular, the Neuroscience Lunch Crew, who have endured the full force of the barrage.

It takes a village to do science, but within the village it takes a family. The good people in my lab have been that family. Thanks to all of you for your inspiration, friendship, rigor, creativity, and hard work.

The idea of writing about human individuality came from a splendid piece by Jeremy Nathans that he wrote for a book of short essays on neuroscience called *Think Tank* that I had the good fortune to edit. Thank you, Jeremy, for sending me on this journey. Soon after I contracted with Basic Books to write this volume in early 2018, two fine books on related topics appeared, Carl Zimmer's *She Has Her Mother's Laugh* and Kevin Mitchell's *Innate.* Thank you to Messrs. Zimmer and Mitchell for your impressive work. It was a joy to read and also served to light a fire under me. The science of human individuality is clearly having a moment in the sun.

Many scientists went out of their way to read and critique sections of the book. For their insightful comments, I thank Seth Blackshaw, Peg McCarthy, Gloria Choi, Paul Breslin, Peter Sterling, and Chip Colwell. Other researchers dug through their files to share images from their key scientific papers. I tip my hat to Nancy Segal, Nicholas Tatonetti, Melissa Hines, and Benoist Schaal.

When the scientists are done, the most valuable insights come from my supersmart lay readers. My deep gratitude goes to the keen minds of Marion Winik and Dena Crosson.

Once again, the publishing pros have made me look better than I really am. Joan Tycko provided clear and compelling illustrations. T. J. Kelleher, Rachel Field, and Liz Dana have edited with a clear eye and a compassionate heart, and Andrew Wylie, Jacqueline Ko, and Luke Ingram at the Wylie Agency have had my back through it all.

Two fine organizations have supported this work. My deepest appreciation goes to the Sloan Foundation Book Program and the Rockefeller Foundation Bellagio Center, the latter of which provided a lovely, collegial, and intellectually engaging environment for writing the final chapter. To all my warm and wonderful Bellagio Center comrades, I raise my glass of Aperol spritz and hope to see you again soon in another inspiring location.

Notes

Prologue

1. If you're interested in learning more about the statistics of online dating, I recommend Christian Rudder's book:

Rudder, C. (2014). *Dataclysm: Love, sex, race, and identity—what our online lives tell us about our offline selves*. New York, NY: Broadway Books.

Chapter One. It Runs in the Family

1. All modern domesticated dogs are descended from an ancient lineage of Eurasian gray wolf that is now extinct. These wolves are the only large carnivore to have been domesticated and are the only species of any type of animal to be domesticated by hunter-gatherers as opposed to agrarian peoples. Comparison of modern wolf and dog DNA suggest that wolf-to-dog domestication was not a onetime event and that domesticated dogs continued to interbreed with wild wolves at several points after their initial domestication. For a recent review of modern dog and wolf genetics, see:

Ostrander, E. A., Wayne, R. K., Freedman, A. H., & Davis, B. W. (2017). Demographic history, selection and functional diversity of the canine genome. *Nature Genetics, 18,* 705–720.

2. Unlike horses, zebras are immune to diseases carried by the African tsetse fly. This knowledge motivated many people to attempt, and fail, to domesticate zebras. If the endeavor had succeeded, it would have been quite useful to African agriculture.

3. It's worthwhile to note that the foxes that were the starting point for the farm-fox breeding experiment, while far from tame, were not entirely wild either. They had been bred in fox farms for many generations and so had already

been slightly selected for human tolerance, if not outright tameness. At the start of Trut and Belyaev's experiment, about 5 percent of male foxes and 20 percent of female foxes were chosen to breed using the tameness criterion.

Trut, L. (1999). Early canid domestication: The farm-fox experiment. *American Scientist, 87,* 160–169.

Lord, K. A., Larson, G., Coppinger, R. P., & Karlsson, E. K. (2019). The history of farm foxes undermines the animal domestication syndrome. *Trends in Ecology & Evolution, 35,* 125–136.

4. For a detailed and engaging story of the Siberian domesticated fox experiments, see:

Dugatkin, L. A., & Trut, L. (2017) *How to tame a fox (and build a dog).* Chicago, IL: University of Chicago Press.

5. Tame fox ordering information is here: https://lkalmanson.com/index.php?option=com_content&view=category&layout=blog&id=19&Itemid=32.

But beware that foxes, even tame ones, are prohibited as pets in some locations, including the states of California, Texas, New York, and Oregon.

6. The quotation is from Wagner, A. (2017, March 31). Why domesticated foxes are genetically fascinating (and terrible pets). PBS *NewsHour.* Retrieved from www.pbs.org/newshour/science/domesticated-foxes-genetically-fascinating-terrible-pets.

7. The phrase "fraternal twins share, on average, 50 percent of their genes" is really an approximation of the more precise but somewhat unwieldy statement that "fraternal twins have a 50 percent chance of inheriting the same version of any particular gene from each parent." It's a subtle but important distinction.

Here, we have made the assumption that each sperm cell involved in fertilization comes from the same father. It is rare, but well-documented, that a woman who has sexual intercourse with two different men during the fertile days of her cycle can conceive fraternal twins that will have different genetic fathers. This phenomenon carries the wonderfully nerdy term of "heteropaternal superfecundication" and is estimated to occur in 0.25 percent of all fraternal twins born to married women in the United States and in 2.4 percent of fraternal twins of parents involved in paternity suits.

James, W. H. (1993). The incidence of superfecundication and of double paternity in the general population. *Acta Geneticae Medicae et Gemellologiae, 42,* 257–262.

Wenk, R. E., Houtz, T., Brooks, M., & Chiafari, F. A. (1992). How frequent is heteropaternal superfecundation? *Acta Geneticae Medicae et Gemellologiae, 41,* 43–47.

Although the rate of heteropaternal superfecundication is higher among cats, dogs, chimpanzees, and some other mammals, the incidence of paternity suits is much lower.

8. Since identical twins are always of the same sex, this study design limits the participating fraternal twins to those that are also of the same sex to avoid a confounding factor.

9. In its simplest form, the equation to estimate heritability (h^2) from the correlation of a trait (r) in identical versus fraternal twins is $h^2 = 2(r_{identical} — r_{fraternal})$. So, for example, if the correlation of a trait is 0.80 in identical twins and 0.50 in fraternal twins, then the heritability of that trait would be $2(0.80—0.50) = 0.60$. In this model, the proportion of the variance attributable to shared environmental

effects (c^2) will be the difference between the identical twin correlation and the heritability: $c^2 = r_{identical} - h^2$.

10. The non-shared environment category also includes measurement error, which is typically small for physical traits but can be more substantial for behavioral traits. For example, the same person, when given a standard personality test or IQ test on two different days, will likely obtain slightly different results.

11. Långström, N., Rahman, Q., Carlström, E., & Lichtenstein, P. (2010). Genetic and environmental effects on same-sex sexual behavior: A population study of twins in Sweden. *Archives of Sexual Behavior, 39,* 75–80.

12. On average, identical twins do share a more similar environment than same-sex fraternal twins. However, it's not clear that it matters very much. Perhaps the best test comes from cases where parents (and the twins, and the community) believed that a twin pair was identical but genetic testing later in life revealed that they were, in fact, fraternal. Or vice versa. One clever study of these misidentified twins showed that parental/community/self beliefs of a twin pair's genetic status had no effect on the similarity of their IQ test scores.

Mathey, A. (1979). Appraisal of parental bias in twin studies: Ascribed zygosity and IQ differences in twins. *Acta Geneticae Medicae et Gemellologiae, 28,* 155–160.

In another study, similar treatment of identical twins by others did not appear to produce more similar behavioral measures. For example, whether twins had the same teachers, dressed alike, or slept in the same room failed to influence their similarity of performance on a standardized scholarship test administered to high school students.

Loehlin, J. C., & Nichols, R. C. (1976). *Heredity, environment and personality: A study of 850 sets of twins.* Austin, TX: University of Texas Press.

This latter finding was largely validated by:

Morris-Yates, A., Andrews, G., Howie, P., & Henderson, S. (1990). Twins: A test of the equal environments assumption. *Acta Physiologica Scandinavica, 81,* 322–326.

13. For many years, it was common practice for adoption agencies in the United States and Europe to place twins and triplets, either identical or fraternal, in separate homes. The presumed rationale for this was that it would be easier to find adoptive parents for a single baby. There is no evidence that this is actually true, but it was a widely held idea. In 2018, the documentary film *Three Identical Strangers* told the story of identical triplet brothers, Edward Galland, David Kellman, and Robert Shafran, who, in 1961, were separated at six months and placed into three different families, one poor, one middle class, and one wealthy. Through a set of coincidences, the triplets reunited at age nineteen. With time, the triplets came to realize that they were separated and placed in these particular families as part of a never-published study on the role of child rearing designed by psychiatrists Peter B. Neubauer and Viola W. Bernard. This study also involved several other adopted identical twin pairs. In my view, the documentary is correct in pointing out the ethical failure of these experiments. However, the filmmakers chose not to reveal that adopted twin separation, completely unrelated to any scientific studies, was a widespread practice in those years, affecting many hundreds of children. They would have the viewer believe that the twins and triplets of the Neubauer/Bernard study were the only siblings ever separated at birth. This does not lessen the culpability of Neubauer and Bernard and their coworkers, but puts it in a more complete historical context.

14. Details of the Jim twins story come from: Hoersten, G. (2015, July 28). Reunited after 39 years. *The Lima News*. Retrieved from www.limaohio.com/features/lifestyle/147776/reunited-after-39-years.

Rawson, R. (1979, May 7). Two Ohio strangers find they're twins at 39—and a dream to psychologists. *People*. Retrieved from https://people.com/archive/two-ohio-strangers-find-theyre-twins-at-39-and-a-dream-to-psychologists-vol-11-no-18/.

15. This anecdote comes from a fine book about the origins, findings, and context of Thomas Bouchard's MISTRA study by Nancy Segal, one of the original MISTRA investigators:

Segal, N. L. (2012). *Born together—reared apart: The landmark Minnesota twin study*. Cambridge, MA: Harvard University Press.

16. One of the well-designed features of MISTRA was its analysis of both fraternal and identical twins raised apart, rather than just the latter. This was an important methodological advance, as some early investigators of twins raised apart recruited only identical twins and in so doing may have unintentionally excluded the less-alike identical sets. The concern is that prescreening may have led prior researchers to believe that identical pairs with notable physical or behavioral differences were fraternal, thereby biasing the pool of participants toward greater similarity. Importantly, twins were accepted into MISTRA without knowledge of their twin type, increasing the probability that the identical pairs represented a broad range of within-pair differences. According to Segal (2012), in the MISTRA trials, twin type was not confirmed by genetic testing until the end of the assessment week.

17. Subsequent work, which is hotly debated, estimated a shared environment contribution to adult population IQ of about 15 percent. That's more than zero but still rather small. Much more on this in chapter 8.

18. Krueger, R. F., Hicks, B. M., & McCue, M. (2001). Altruism and antisocial behavior: Independent tendencies, unique personality correlates, distinct etiologies. *Psychological Science, 12,* 397–402.

19. Lejarraga, T., Frey, R., Schnitzlein, D. D., & Hertwig, R. (2019). No effect of birth order on adult risk taking. *Proceedings of the National Academy of Sciences of the USA, 116,* 6019–6024.

Damian, R. I., & Roberts, B. W. (2015). The associations of birth order with personality and intelligence in a representative sample of US high school students. *Journal of Research in Personality, 58,* 96–105.

Botzet, L., Rohrer, J. M., & Arslan, R. C. (2018). Effects of birth order on intelligence, educational attainment, personality, and risk aversion in an Indonesian sample. *PsyArXiv*. doi:10.31234/osf.io/5387k.

20. Polderman, T. J. C., Benyamin, B., de Leeuw, C. A., Sullivan, P. F., von Bochoven, A., Visscher, P. M., & Posthuma, D. (2015). Meta-analysis of the heritability of human traits based on fifty years of twin studies. *Nature Genetics, 47,* 702–709.

21. Quotation from Miller, P. (2012, January). A thing or two about twins. *National Geographic*. Retrieved from www.nationalgeographic.com/magazine/2012/01/identical-twins-science-dna-portraits/.

22. Interestingly, malnutrition is not just a problem of poverty. There is epidemiological evidence suggesting that contemporary Japanese women, by

consuming fewer calories in their desire to stay slim during pregnancy, are stunting the growth of their children.

Normile, D. (2018). Staying slim during pregnancy carries a price. *Science, 361,* 440.

23. Nisbett, R. E., et al. (2012). Intelligence: New findings and theoretical developments. *American Psychologist, 6,* 130–159.

24. DNA, or deoxyribonucleic acid, is a chain of repeated chemical units called nucleotides (bases). Nucleotides are molecules that combine an amino acid with a sugar. There are four nucleotides: adenine, guanine, thymine, and cytosine (A, G, T, C). DNA exists as two chains that spiral around each other, creating the famous double helix structure. If there is a C on one chain, it pairs with a G on the other; if there is a T, it pairs with an A. In this way, both chains of the double helix contain the same information in mirror images. The key insight is that groups of three nucleotides form the genetic alphabet: each triplet of DNA codes for a specific amino acid, chains of which form proteins.

When a particular protein is made in a cell, the relevant portion of the DNA code also instructs where a gene starts and the strands unzip. One side of the DNA strand then serves as a template to make a complementary strand of ribonucleic acid (RNA). Free nucleotides will line up along the unzipped DNA strand, pairing C to G and T to U (uracil), to form an RNA strand. This chain of messenger RNA then peels away from the DNA strand. In this way, the sequence of bases in the messenger RNA strand mirrors the sequence of bases on the DNA strand. The strand of messenger RNA then leaves the nucleus and enters the cytoplasm of the cell, where it stops. Messenger RNA then interacts with protein complexes called ribosomes and the information is translated when amino acids line up along the messenger RNA strand to form a protein.

Almost all genetic information acts by instructing RNA, and hence protein production and the function of organisms is ultimately determined by the particular types and amounts of various proteins that different types of cells make. There are also some noncoding RNAs that work independently of instructing protein synthesis, but I won't go into those here.

25. Some plants have a large number of genes because, during evolution, they have undergone whole-genome duplication once, twice, or even three times.

26. PKU is a serious disease, but it is easily treated by holding to a diet low in phenylalanine. This is why newborns are routinely screened for it.

27. There are two ways in which a mutation of a single nucleotide in a gene can fail to have a functional effect. The first is called a silent mutation. That's when two different triplets of DNA encode the same amino acid. For example, if the triplet AAA were altered to become AAG, the same amino acid—lysine—will be incorporated into the chain of amino acids that forms the protein. In the second case, called a conservative substitution, changing a single nucleotide does change the amino acid—say, from a glutamate to an aspartate—but that particular change at that particular location has no effect on the protein's function.

28. There are some notable exceptions. There are some genes where even though you inherit two copies, only the one from your mother or from your father is active. This process is called epigenetic imprinting. For example, a gene called *UBE3A*, which is important for the development and function of the nervous system, is only expressed from the maternal allele. When there are

mutations in the maternal allele (or problems with expressing it), this can result in a disease of the nervous system called Angelman syndrome. Other exceptions are genes expressed on the X or Y sex chromosomes. For example, loss-of-function mutations in a gene on the X chromosome can cause diseases in males whose other chromosome is a Y, while females, who typically have two X chromosomes, will usually lack the disease, due to having one normal allele (unless they are unlucky enough to inherit mutant alleles from both parents). An example of this is red/green color blindness, which affects about 8 percent of males and 0.5 percent of females in a northern European population.

29. If you like to complain about your job, just remember that it could be worse. You could be an international earwax harvester.

30. Yoshiura, K., et al. (2006). A SNP in the *ABCC11* gene in the determinant of human earwax type. *Nature Genetics, 38,* 324–330.

Nakano, M., Miwa, N., Hirano, A., Yoshiura, K., & Niikawa, N. (2009). A strong association of axillary osmidrosis with the wet earwax type determined by genotyping of the *ABCC11* gene. *BMC Genetics, 10,* 42.

31. The rest of your body mostly has a different kind of sweat gland—eccrine glands—that secretes a watery, salty sweat that does not provide metabolites for the extra-stinky bacteria of the pits and crotch. The armpits support a dense population of *Corynebacterium* and *Staphylococcus*, which appear to be the funky bacterial culprits.

32. Rodriguez, S., Steer, C. D., Farrow, A., Golding, J., & Day, I. N. (2013). Dependence of deodorant usage on ABC11 genotype: Scope for personalized genetics in personal hygiene. *Journal of Investigative Dermatology, 133,* 1760–1767.

33. Also, some of the gene variation underlying height was actually found in stretches of DNA between genes.

Wood, A. R., et al. (2014). Defining the role of common variation in the genomic and biological architecture of adult human height. *Nature Genetics, 46,* 1173–1186.

Marouli, E., et al. (2017). Rare and low-frequency coding variants alter human height. *Nature, 542,* 186–190.

34. Knowing that about 50 percent of the variation in a particular dimension of personality, like extraversion, can be explained by genes is only the beginning. What particular genes influence extraversion and how do they do so? And how do such gene variants interact with experience, both social and nonsocial? Here, our understanding becomes very limited. Let's take one recent study as an example. Investigators from University of California, San Diego, and several other universities performed standard five-factor OCEAN personality tests and collected DNA from about two hundred thousand people. Then they performed statistical tests to try to find places in the human genome where genetic variation predicted a portion of any one of the five OCEAN personality measures. This type of investigation is called a genome-wide association study and has the advantage that it doesn't start out with any preconceived ideas about which genes might be important. They found several interesting correlations, one of which was a relationship between extraversion and variation in a gene called *WSCD2*. Does this mean that *WCSD2* is the extraversion gene? Absolutely not! Variation in *WSCD2* accounts for less than 10 percent of the variation in extraversion in the sampled population. Even if this finding were to be replicated, and it may well

be, *WSCD2* will only be a small part of the entire genetic contribution. It would be nice if *WSCD2* had some relationship to specific circuits in the brain that we think might be related to extraversion, like the dopamine- and serotonin-using neurons. At present, there's no such link. The protein made from the *WSCD2* gene is not enriched in these types of neuron. In fact, it is present in the thyroid gland at levels that are even higher than the brain.

Lo, M. T., et al., (2017). Genome-wide analyses for personality traits identify six genomic loci and show correlations with psychiatric disorders. *Nature Genetics, 49,* 152–156.

35. VonHoldt, B. M., et al. (2017). Structural variants in genes associated with human Williams-Beuren syndrome underlie stereotypical hypersocialibility in domestic dogs. *Science Advances, 3.* doi:10.1126/sciadv.1700398.

Chapter Two. Are You Experienced?

1. This phrase has a tradition that can be traced back to medieval epic poetry, including *Perceval, the Story of the Grail,* written by Chrétien de Troyes in Old French around 1180.

2. There are a few short-lived cells, like red blood cells and blood platelets, that don't have a nucleus and hence don't have nuclear DNA, but they are the exception.

3. It's important to realize that there are many different types of neuron and each type of neuron will have a somewhat different pattern of gene expression. Often, we can relate aspects of this pattern to the function of the neuron. For example, neurons that have to fire electrical signals very fast, say one hundred times per second or more, will express genes for ion channels that open very quickly, allowing for rapid electrical signaling, while neurons that tend to fire more slowly will not, or will express those genes at much lower levels. At present there are several efforts underway to determine how many different "flavors" of neuron there are, as defined by patterns of gene expression. You can read about one of these projects at http://celltypes.brain-map.org/.

4. Specifically, it's the cytosine (C) and adenine (A) nucleotides in the chain of DNA that are methylated.

5. As always, the more one digs in, the more details emerge. While most transcription factors regulate genes by binding near a gene's transcription start site, others bind much further away but work by bending the DNA in a loop back toward the start site. Another complicating factor is called alternative splicing, in which a single gene has segments within it that can be included or not in the transcribed gene produced (reflected in the sequence of the mRNA). By having many of these alternative splicing sites, some genes can give rise to hundreds or even thousands of different modular proteins.

6. The term "epigenetics" is having a bit of a moment in popular culture and in the realm of pseudoscience. While epigenetics—the regulation of gene expression that does not change the sequence of the DNA—is real, the notion that experience of your ancestors, particularly their trauma, can be transmitted to you across generations remains unproven and has been the basis of a lot of junk science. There are plenty of people on the internet

and elsewhere who are happy to sell you bogus treatments to "clear your epigenetics going back nine generations." My suggestion: tell those quacks to piss off.

7. The sweating Japanese soldiers story comes from this splendid book about lasting environmental effects produced by perturbations in the fetal or early postnatal stages:

Gluckman, P., & Hanson, M. (2005). *The fetal matrix: Evolution, development and disease* (pp. 7–8). Cambridge, UK: Cambridge University Press.

8. Valenzuela, N., & Lance, V. (2004). (Eds.). *Temperature-dependent sex determination in vertebrates*. Washington, DC: Smithsonian Institution Press.

Lang, J. W., & Andrews, H. V. (1994). Temperature-dependent sex determination in crocodilians. *The Journal of Experimental Zoology, 270,* 28–44.

It's not clear if or how temperature-dependent sex determination provides an evolutionary advantage to those fish and reptile species that have it.

Recently, it has been shown that, in the snapping turtle, a gene called *CIRBP* is differentially expressed in male- and female-producing temperatures and influences the path for primordial tissues to develop into either ovaries or testes.

Schroeder, A. L., Metzger, K. J., Miller, A., & Rhen, T. (2016). A novel candidate gene for temperature-dependent sex determination in the common snapping turtle. *Genetics, 203,* 557–571.

9. Lee, T. M., & Zucker, I. (1988). Vole infant development is influenced perinatally by maternal photoperiodic history. *American Journal of Physiology, 255,* R831–R838.

10. Boland, M. R., et al. (2015). Birth month affects lifetime disease risk: A phenome-wide method. *Journal of the American Medical Informatics Association, 22,* 1042–1053.

This study followed others that had reported birth month associations with longevity and reproductive performance, as well as many diseases including myopia, multiple sclerosis, and atherosclerosis. Of the fifty-five conditions associated with birth month here, nineteen had been previously reported in the literature by 2015. In fact, the variation in these birth-month-disease correlations across locations is interesting in its own right. For example, there was a two-month shift in the peak association between birth month and asthma between this New York City study and one based in Denmark that is well explained by a shift in the peak sunshine curve between the two sites.

Korsgaard, J., & Dahl, R. (1983). Sensitivity to house dust mite and grass pollen in adults. Influence of the month of birth. *Clinical Allergy, 13,* 529–535.

11. Disanto, G., et al. (2012). Month of birth, vitamin D and risk of immune-mediated disease: A case control study. *BMC Medicine, 10,* 69.

12. Boland, M. R., et al. (2018). Uncovering exposures responsible for birth season-disease effects: A global study. *Journal of the American Medical Informatics Association, 25,* 275–288.

A subsequent study, restricted to the United States, showed a similar result:

Layton, T. J., Barnett, M. L., Hicks, T. R., & Jena, A. B. (2018). Attention deficit-hyperactivity disorder and month of school enrollment. *New England Journal of Medicine, 379,* 2122–2130.

13. Soreff, S. M., & Bazemore, P. H. (2008). The forgotten flu. *Behavioral Healthcare, 28,* 12–15.

The 1918 flu was made worse by rampant misinformation. Some claimed that the flu was a German weapon, brought to American shores by U-boats. Others, predictably, blamed immigrants—many Denver residents, for example, focused on Italians as the source. Authorities often did little to help. The city of Philadelphia had one of the highest flu mortality rates, likely due to the refusal of leaders to cancel large public events such as a citywide parade that drew two hundred thousand spectators.

14. Nearly every family was affected by the 1918 pandemic flu: Woodrow Wilson, Mary Pickford, and Walt Disney survived. The French surrealist poet Guillaume Apollinaire, the American suffragist Phoebe Hearst, and the Austrian painter Egon Schiele were not so fortunate and died from their flu infections.

15. Mazumder, B., Almond, D., Park, K., Crimmins, E. M., & Finch, C. F. (2010). Lingering prenatal effects of the 1918 influenza pandemic on cardiovascular disease. *Journal of Developmental Origins of Health and Disease, 1,* 26–34.

16. Brown, A. S., et al. (2004). Serologic evidence of prenatal influenza in the etiology of schizophrenia. *Archives of General Psychiatry, 61,* 774–780.

This study didn't just rely on mothers' reports of influenza during pregnancy, but rather validated them by measuring antibodies against influenza in archived maternal blood samples. The biggest effects were seen with influenza infection during the first trimester of pregnancy.

17. Lee, B. K., et al. (2015). Maternal hospitalization with infection during pregnancy and risk of autism spectrum disorders. *Brain, Behavior and Immunity, 44,* 100–105.

18. Smith, S. E. P., Li, J., Garbett, K., Mirnics, K., & Patterson, P. H. (2007). Maternal immune activation alters fetal brain development through interleukin-6. *Journal of Neuroscience, 27,* 10695–10702.

19. Choi, G. B., et al. (2016). The maternal interleukin-17a pathway in mice promotes autism-like phenotypes in offspring. *Science, 351,* 933–939.

The specific immune cell in the mother that secretes IL-17a is white blood cell called a T helper 17 lymphocyte. To interfere with IL-17a function, an inactivating antibody against IL-17a was used.

Interestingly, maternal infection produced offspring with a range of disrupted cortical patches. Those mice that grew up with the largest area of cortical patches also showed the strongest autism-like behaviors.

Yim, Y. S., et al. (2017). Reversing behavioral abnormalities in mice exposed to maternal inflammation. *Nature, 549,* 482–487.

20. Stoner, R., et al. (2014). Patches of disorganization in the neocortex of children with autism. *New England Journal of Medicine, 370,* 1209–1219.

Al-Aayadhi, L. Y., & Mostafa, G. A. (2012). Elevated levels of interleukin 17a in children with autism. *Journal of Neuroinflammation, 9,* 158.

21. Lammert, C. R., et al. (2018). Critical roles for microbiota-mediated regulation of the immune system in a prenatal immune activation model of autism. *Journal of Immunology*, 1701755.

22. Kim, S., et al. (2017). Maternal gut bacteria promote neurodevelopmental abnormalities in mouse offspring. *Nature, 549,* 528–532.

In addition to infecting SFB-free mice with SFB from other mice, they also tried strains of bacteria that normally live in the gut of humans and cause T helper 17 cell differentiation. They were similarly able to support maternal-infection-evoked autism-like cortical patches and behaviors.

23. Turecki, G., & Meaney, M. J. (2016). Effects of social environment and stress on glucocorticoid receptor gene methylation: A systematic review. *Biological Psychiatry, 79,* 87–96.

Interestingly, a similar effect may be seen in rats. Most rat mothers spend a lot of time licking and snuggling their pups. Those that don't do this have pups with increased methylation of the glucocorticoid receptor gene's regulatory region and enhanced CRH-mediated stress reactivity. Behaviorally, this is manifested as increased anxiety, decreased exploratory behavior, and reduced willingness to try new foods.

24. Streisand, B. (2018, March 2). Barbra Streisand explains: Why I cloned my dog. *The New York Times.* Retrieved from www.nytimes.com/2018/03/02/style/barbra-streisand-cloned-her-dog.html.

25. Medland, S. E., Loesch, D. Z., Mdzewski, B., Zhu, G., Montgomery, G. W., & Martin, N. G. (2007). Linkage analysis of a model quantitative trait in humans: Finger ridge count shows significant multivariate linkage to 5q14.1. *PLOS Genetics, 3,* 1736–1744.

26. Pinc, L., Bartoš, L., Restová, A., & Kotrba, R. (2011). Dogs discriminate identical twins. *PLOS One, 6,* 1–4.

27. Lykken, D. T., & Tellegen, A. (1993). Is human mating adventitious or the result of lawful choice? A twin study of mate selection. *Journal of Personality and Social Psychology, 65,* 56–68.

In this study, which was limited to heterosexual couples, only 7 percent of women and 13 percent of men said that they might have fallen in love with their spouse's identical twin brother or sister. The authors write, charmingly, that "it is romantic infatuation that commonly determines the final choice from a broad field of potential eligibles and that this phenomenon is inherently random."

28. This has not stopped scientists who should know better from making overly deterministic claims about the genome. For example, Robert Plomin, a behavioral geneticist from Kings College, London, has a new book in which he calls DNA a "fortune teller" that is "100 percent reliable." That may hold for the trait of earwax type, but it fails for every human behavioral trait and nearly every human structural trait.

Plomin, R. (2018). *Blueprint: How DNA makes us who we are.* Cambridge, MA: MIT Press.

29. Mitchell, K. J. (2018). *Innate: How the wiring of our brains shapes who we are.* Princeton, NJ: Princeton University Press.

Mitchell encapsulates the role of intrinsic variation in the development of the nervous system with the observation that, even using the same recipe, "you can't bake the same cake twice."

30. Fraga, M. F., et al. (2005). Epigenetic differences arise during the lifetime of monozygotic twins. *Proceedings of the National Academy of Sciences of the USA, 102,* 10604–10609.

31. Lodato, M. A., et al. (2015). Somatic mutation in single human neurons tracks developmental and transcriptional history. *Science, 350,* 94–98.

32. While most neurons in the mammalian brain have stopped dividing (they are called post-mitotic), there are two restricted locations in the brain where neuronal precursors continue to divide throughout life: the dentate gyrus of the hippocampus, a structure involved in spatial learning and memory, and the subventricular zone, which produces certain neurons of the olfactory bulb. While it

seems clear that this restricted neurogenesis occurs in rats and mice, whether it also occurs in adult humans remains unresolved.

Kuhn, H. G. (2018). Adult hippocampal neurogenesis: A coming-of-age story. *Journal of Neuroscience, 38,* 10401–10410.

33. Single-nucleotide somatic mutations don't occur entirely randomly. They are more prevalent in regions of the DNA that are actively being read out (transcribed) to instruct the production of proteins. There seems to be something about the process of transcription that renders the DNA more subject to mutation. It should also be noted that there are other ways for spontaneous somatic mutation to occur. One of these involves a segment of DNA called an L1 retrotransposon that "jumps" around the genome, potentially creating havoc—or, rarely, something new and good—wherever it lands. For a nice review of somatic mosaicism in the brain, see:

Paquola, A. C. M., Erwin, J. A., & Gage, F. H. (2017). Insights into the role of somatic mosaicism in the brain. *Current Opinion in Systems Biology, 1,* 90–94.

34. When these large mutations impair genes that control cell division, rampant cell division can occur and cancer results. Most cancer cells in a tumor are genetically identical, suggesting that they originate from mutation in a single cell. Not all cancer-causing mutations occur randomly. Some are produced by viral infection, like cervical cancer from exposure to human papilloma virus. Others are triggered by exposure to things that cause mutations, like UV light (the origin of many skin cancers), X-rays, or chemicals that interact with DNA, like certain compounds found in cigarette smoke.

35. Poduri, A., et al. (2012). Somatic activation of AKT3 causes hemispheric developmental brain malformation. *Neuron, 74,* 41–48.

36. Dunsford, I., Bowley, C. C., Hutchison, A. M., Thompson, J. S., Sanger, R., & Race, R. R. (1953). A human blood-group chimera. *British Medical Journal, 11,* 81.

37. Martin, A. (2007). "Incongruous juxtapositions": The chimaera and Mrs. McK. *Endeavour, 31,* 99–103.

38. Chimerism is different than somatic mosaicism, where various cells in the body have different genomes but all derive from the same individual.

39. Gammill, H. S., & Nelson, J. L. (2010). Naturally acquired microchimerism. *International Journal of Developmental Biology, 54,* 531–543.

40. Chan, W. F. N., et al. (2012). Male microchemerism in the human female brain. *PLOS One, 7,* e45592.

In this study and several others, fetal-to-maternal chimerism was assessed by measuring the presence of male DNA in the mother's brain. This is not because male fetal cells have a special role in invading the maternal body, but just because it makes the measurement easy to do, as male DNA is not typically present in mothers otherwise.

41. Of course, this realization complicates how we think about surrogate pregnancy, where maternal cells will transfer and linger in the child and fetal cells will transfer and linger in the surrogate mother.

42. Bianchi, D. W., & Khosrotehrani, K. (2005). Multi-lineage potential of fetal cells in maternal tissue: A legacy in reverse. *Journal of Cell Science, 18,* 1559–1563.

43. Bianchi, D. W. (2007). Fetomaternal cell trafficking: A story that begins with prenatal diagnosis and may end with stem cell therapy. *Journal of Pediatric Surgery, 42,* 12–18.

44. Pembrey, M. E., et al. (2006). Sex-specific, male-line transgenerational responses in humans. *European Journal of Human Genetics, 14,* 159–166.

Bygren, L. O., et al. (2014). Changes in paternal grandmother's early food supply influenced cardiovascular mortality of the female grandchildren. *BMC Genetics, 30,* 173–195.

45. Kevin Mitchell provides a useful analysis of the shortcomings of these studies on his blog:

Mitchell K. (2018, May 29). Grandma's trauma—a critical appraisal of the evidence for transgenerational epigenetic inheritance in humans [Blog post]. Retrieved from www.wiringthebrain.com/2018/05/grandmas-trauma-critical-appraisal-of.html.

Mitchell K. (2018, July 22). Calibrating scientific skepticism—a wider look at the field of transgenerational epigenetics [Blog post]. Retrieved from www.wiringthebrain.com/2018/07/calibrating-scientific-skepticism-wider.html.

46. Hackett, J. A., et al. (2013). Germline DNA demethylation dynamics and imprint erasure through 5-hydroxymethylcytosine. *Science, 339,* 448–452.

47. Miska, E. A., & Ferguson-Smith, A. C. (2016). Transgenerational inheritance: Models and mechanisms of non-DNA sequence-based inheritance. *Science, 354,* 59–63.

48. Buchanan, S. M., Kain, J. S., & de Bivort, B. L. (2015). Neuronal control of locomotor handedness in *Drosophila. Proceedings of the National Academy of Sciences of the USA, 112,* 6700–6705.

49. This is not just a fruit fly trick. Consistent behavioral variability in the decision to drop or cling is also seen in genetically identical pea aphids when confronted with a potential predator.

Schuett, W., et al. (2011). "Personality" variation in a clonal insect: The pea aphid *Acryrthosiphon pisum. Developmental Psychobiology, 53,* 631–640.

50. This process is called a diversified bet hedging strategy.

Honneger, K., & de Bivort, B. (2018). Stochasticity, individuality and behavior. *Current Biology, 28,* R1–R5.

Chapter Three.
I Forgot to Remember to Forget You

1. Carlson, P. (1997, March 23). In all the speculation and spin surrounding the Oklahoma City Bombing, John Doe 2 has become a legend—the central figure in countless conspiracy theories that attempt to explain an incomprehensible horror. Did he ever really exist? *The Washington Post.* Retrieved from www.washingtonpost.com/archive/lifestyle/magazine/1997/03/23/in-all-the-speculation-and-spin-surrounding-the-oklahoma-city-bombing-john-doe-2-has-become-a-legend-the-central-figure-in-countless-conspiracy-theories-that-attempt-to-explain-an-incomprehensible-horror-did-he-ever-really-exist/04329b31-ddfa-4ddb-9404-b9944ceca2b3/.

2. Mistaken eyewitness identifications resulted in approximately 71 percent of the more than 350 wrongful convictions in the United States overturned by post-conviction DNA evidence.

Eyewitness identification reform. (n.d.). Retrieved from www.innocenceproject.org/eyewitness-identification-reform/.

3. A better way to query an eyewitness is to present each person in the lineup separately, in a random order, for a yes-or-no answer, without telling the eyewitness in advance how many such determinations they will have to make. Furthermore, the person running the lineup should be unaware of which person is believed by the police to be the likely suspect in order to avoid giving subtle cues with their voice or gestures. This practice, called blind sequential lineup, is now routine among police in many jurisdictions in the United States and Europe, and is almost certainly reducing the rate of false convictions. It's disheartening that this protocol is not the universal accepted legal standard for eyewitness identification.

4. Schacter, D. (2001). *The seven sins of memory: How the mind forgets and remembers*. Boston, MA: Mariner Books.

5. Nigro, G., & Neisser, U. (1983). Point of view in personal memories. *Cognitive Psychology, 15*, 467–482.

Robinson, J. A., & Swanson, K. L. (1993). Field and observer modes of remembering. *Memory, 1,* 169–184.

6. There are several lines of evidence implicating impaired function of the frontal lobes with source amnesia, particularly in the elderly.

Craik, F. I. M., Morris, L. W., Morris, R. G., & Loewen, E. R. (1990). Relations between source amnesia and frontal lobe functioning in older adults. *Psychology and Aging, 5*, 148–151.

Dywan, J., Segalowitz, S. J., & Williamson, L. (1994). Source monitoring during name recognition in older adults: Psychometric and electrophysiological correlates. *Psychology and Aging, 9,* 568–577.

7. There are some forms of implicit memory that are formed from single events. For example, if you eat food that then causes you to become ill, you will develop a lasting subconscious aversion to the sight and smell of that food from a single unpleasant experience. Importantly, this subconscious conditioned taste aversion is accompanied by separate explicit memories of the event.

8. Damage to the medial temporal lobes can come about through stroke, infections, chronic abuse of drugs or alcohol, or, in one famous case—the oft-discussed patient Henry Molaison (known for years as H. M.)—surgical resection of the temporal lobes to control otherwise intractable epilepsy. The relevant structures in the temporal lobe for producing anterograde amnesia appear to be the rhinal and parahippocampal cortex and the hippocampus. Similar profound anterograde and delimited retrograde amnesia can be produced by damaging these regions in laboratory animals.

9. The classic paper showing that acquisition in mirror reading is preserved in temporal lobe amnesia patients is:

Cohen, N. J., & Squire, L. R. (1980). Preserved learning and retention of pattern analyzing skill in amnesia: Dissociation of knowing how and knowing that. *Science, 210,* 207–209.

10. For a nice review of human memory research with particular attention to lessons from amnesiac patients, see:

Squire, L. R., & Wixted, J. T. (2011). The cognitive neuroscience of human memory since H.M. *Annual Review of Neuroscience, 34,* 259–288.

11. Sechenov, I. (1863). Refleksy golovnogo mozga. *Meditsinsky Vestnik*. In English: Sechenov, I. (1965). *Reflexes of the Brain* (S. Belsky, Trans.). Cambridge, MA: MIT Press.

12. For the orienting response to be elicited, the stimulus must be novel but not threatening. If, for example, a very loud sound is used, this will elicit a defensive turning-away or flinch reflex rather than an orienting reflex.

13. Pribram, K. H. (1969). The neurophysiology of remembering. *Scientific American, 220,* 73–87.

14. Haith, A. M. (2018). Almost everything you do is a habit. In D. J. Linden, (Ed.), *Think tank: Forty neuroscientists explore the biological roots of human experience.* New Haven, CT: Yale University Press.

This short chapter is a lovely read. And I'm not just saying that because I edited the volume in which it appears.

15. Woodruff-Pak, D. S. (1993). Eyeblink classical conditioning in H.M.: Delay and trace paradigms. *Behavioral Neuroscience, 107,* 911–925.

16. Many other brain regions are implicated in implicit memory including the striatum, the cerebellum, the amygdala, and several regions of the neocortex.

17. The evidence implicating these brain regions and their interconnection comes from several different types of experiment, including analysis of working memory in humans with brain damage in various regions, examination of laboratory animals in which specific brain regions have been carefully destroyed or inactivated, and, most compellingly, the combination of rapid reversible inactivation of specific brain regions with recordings of the activity of individual neurons in those brain regions.

For a classic paper that began to point the way to frontal cortex see:

Fuster, J. M., & Alexander, G. E. (1971). Neuron activity related to short-term memory. *Science, 173,* 652–654.

For modern papers that have fleshed out the long-range reverberatory loops connecting the frontal cortex with other brain regions, see:

Guo, Z. V., et al. (2017). Maintenance of persistent activity in a frontal thalamocortical loop. *Nature, 545,* 181–186.

Gao, Z., Davis, C., Thomas, A. M., Economo, M. N., Abrego, A. M., Svoboda, K., De Zeeuw, C. I., & Li, N. (2018). A cortico-cerebellar loop for motor planning. *Nature, 563,* 113–116.

18. This is a super-simplified account. Some synapses are received on dendrites, others on the cell body, and yet others on the axon. Some neurotransmitter receptors increase the probability of spike firing in the neuron on which they are received (excitation), others reduce the probability of spike firing (inhibition), and still others have complex actions that aren't either excitation or inhibition (neuromodulation). For a somewhat more detailed explanation, see:

Linden, D. J. (2007). *The accidental mind: How brain evolution has given us love, memory, dreams, and God* (pp. 28–49). Cambridge, MA: Harvard University Press.

19. The famous and impressive arithmetic: There are estimated to be about one hundred billion neurons in the brain. So, with an average of five thousand synapses per neuron, that comes out to about five hundred trillion synapses. By comparison, there are estimated to be between one hundred and four hundred billion stars in our galaxy.

20. Memory storage does not typically require gene expression for the first hour or so after experience, but if drugs are given that block the readout for genes to make RNA and then proteins, this produces long-term memory deficits.

21. In humans, we can't take biopsy samples or implant electrodes in the brains of taxi drivers, so we're limited to noninvasive measurements, like the size of various brain regions as measured using magnetic resonance imaging (MRI) machines. Here is the initial MRI study:

Maguire, E. A., Gadian, D. G., Johnsrude, I. S., Good, C. D., Ashburner, J., Frackowiak, R. S. J., & Frith, C. D. (2000). Navigation-related structural changes in the hippocampi of taxi drivers. *Proceedings of the National Academy of Sciences of the USA, 97,* 4398–4403.

Another interesting finding of this study was that the degree of enlargement of the posterior hippocampus was positively correlated with the total time spent learning to be and then serving as a London taxi driver.

22. Woollett, K., & Maguire, E. A. (2011). Acquiring "the Knowledge" of London's layout drives structural brain changes. *Current Biology, 21,* 2109–2114.

23. Woolett, K., Spiers, H. J., & Maguire, E. A. (2009). Talent in the taxi: A model system for exploring expertise. *Philosophical Transactions of the Royal Society, Series B, 364,* 1407–1416.

24. The posterior hippocampal increases associated with studying were confined to the left side of the brain for reasons that are not entirely clear.

Draganski, B., Gaser, C., Kempermann, G., Kuhn, H. G., Winkler, J., Büchel, C., & May, A. (2006). Temporal and spatial dynamics of brain structure changes during extensive learning. *Journal of Neuroscience, 26,* 6314–6317.

25. Woollett, K., Glensman, J., & Maguire, E. A. (2008). Non-spatial expertise and hippocampal gray matter volume in humans. *Hippocampus, 18,* 981–984.

26. The increase in volume produced by juggling was small, about 3 percent, but statistically significant. Notably, it was confined to the gray matter of the neocortex, the layer containing cell bodies, dendrites, and unmyelinated axons, but fewer myelinated axons (which mostly run in the white matter). Unlike the London taxi drivers and the German medical students, the jugglers did not show reductions in the volume of any brain regions, adjacent or otherwise.

Draganski, B., Gaser, C., Busch, V., Schuierer, G., Bogdhan, U., & May, A. (2004). Changes in grey matter induced by training. *Nature, 427,* 311–312.

27. Van Dyck, L. I., & Morrow, E. M. (2017) Genetic control of postnatal human brain growth. *Current Opinion in Neurobiology, 30,* 114–124.

28. For a nice review of the state of the debate on human neurogenesis, see:

Snyder, J. S. (2018). Questioning human neurogenesis. *Nature, 555,* 315–316.

Another way to increase the volume of a brain region is to increase the degree of wrapping of axons with the insulating protein myelin. This can occur in both gray and white matter layers of the brain.

29. The longer a string player played, the larger the increase in left-hand representation would be. This finding suggests but does not prove that the act of sustained string instrument playing caused the enlargement of the left-hand representation in the somatosensory cortex. It could be that those people born with larger left-hand representations are more likely to take up and succeed at playing string instruments. This is why prospective studies that start before training—like the London taxi drivers, the jugglers, and the German medical students—are so useful.

Elbert, T., Pantev, C., Weinbruch, C., Rockstroh, B., & Taub, E. (1995). Increased cortical representation of the fingers of the left hand in string players. *Science, 270,* 305–307.

Chapter Four. Sexual Self

1. As is often the case in biology, there are caveats. One of the key targets of the *SRY* gene product is another transcription factor called Sox9. This means that loss-of-function mutations in the *SOX9* gene can also block testicular development. And certain people with XX chromosomes develop testes even in the absence of the *SRY* gene (maybe some of them have a gain-of-function mutation in a key *SRY* target like *SOX9*). This is one of the reasons why a test for the *SRY* gene product was abandoned for female athlete screening.

2. Some people prefer the term disorders of sexual development (DSD).

3. In androgen-insensitivity syndrome in XY individuals, the surge of testosterone that accompanies puberty cannot act on androgen receptors to produce secondary sexual characteristics. Instead, this testosterone is converted into estrogen by aromatase enzymes and binds functional estrogen receptors to produce female-typical secondary sex characteristics.

For a thoughtful and humane TED talk by Emily Quinn, who has androgen insensitivity syndrome, see: Quinn, E. (Presenter). (2018). *The way we think about biological sex is wrong* [Video File]. Retrieved from www.ted.com/talks/emily_quinn_the_way_we_think_about_biological_sex_is_wrong/transcript?language=en.

Her talk begins:

"I have a vagina. Just thought you should know. That might not come as a surprise to some of you. I look like a woman. I'm dressed like one, I guess. The thing is, I also have balls. And it does take a lot of nerve to come up here and talk to you about my genitalia. Just a little. But I'm not talking about bravery or courage. I mean literally—I have balls. Right here, right where a lot of you have ovaries. I'm not male or female. I'm intersex."

4. Hughes, I. A. (2002). Intersex. *BJU International, 90,* 769–776.

Okeigwe, I., & Kuohung, W. (2014). 5-alpha reductase deficiency: A 40-year retrospective review. *Current Opinion in Endocrinology, Diabetes and Obesity, 21,* 483–487.

5. In about 95 percent of cases, congenital adrenal hyperplasia results from mutations in the gene *CYP21A2* that directs expression of the enzyme 21-hydroxylase. 21-hydroxylase interferes with cortisol production, and so cortisol precursor chemicals build up and are fed into the androgen-producing pathway instead. The end result is overproduction of fetal androgens.

6. In rare cases, XX people are masculinized during development, not by androgen secretion from their own adrenal glands, but rather by the disordered adrenal glands of their mother acting across the placenta.

Morris, L. F., Park, S., Daskivich, T., Churchill, B. M., Rao, C. V., Lei, Z., Martinez, D. S., & Yeh, M. W. (2011). Virilization of a female infant by a maternal adrenocortical carcinoma. *Endocrine Practice, 17,* e26–e31.

7. Estimates for the incidence of intersex conditions observed at birth range from 0.018 percent to as high as 1.7 percent if chromosomal anomalies like Klinefelter's syndrome (XXY) and Turner's syndrome (XO, XXX, and XYY) are included. In my view, only a fraction of individuals with Klinefelter's syndrome should be included, because they can have obvious sex-related traits like breast enlargement and testicular atrophy, but most don't and they mostly identify in a cisgender fashion. In addition, it's not clear to me that the other

chromosomal anomalies should be counted in calculating the rate of intersex conditions. For example, Turner's syndrome girls have normal external genitalia and some have attenuated secondary sexual characteristics (and nearly all unambiguously identify with the female sex). Using these criteria, the incidence of intersex conditions is about 0.03 percent.

For a useful definition of intersex-related terms and information about advocacy for intersex youth, see: Intersex definitions (n.d.). Retrieved from https://interactadvocates.org/intersex-definitions/.

8. Diamond M., & Sigmundson, H. K. (1997). Sex reassignment at birth: Long-term review and clinical implications. *Archives of Pediatric and Adolescent Medicine, 151,* 298–304.

9. Ritchie, R., Reynard, J., & Lewis, T. (2008). Intersex and the Olympic games. *Journal of the Royal Society of Medicine, 101,* 395–399.

Ha, N. Q., et al. (2014). Hurdling over sex? Sport, science and equity. *Archives of Sexual Behavior, 43,* 1035–1042.

10. Carlson, A. (2005). Suspect sex. *Lancet, 366,* S39–S40.

11. Martínez-Patiño, M. J. (2005). A woman tried and tested. *Lancet, 366,* S38.

12. Ferguson-Smith, M. A., & Bavington, L. D. (2014). Natural selection for genetic variants in sport: The role of Y chromosome genes in elite female athletes with 46, XY DSD. *Sports Medicine, 44,* 1629–1634.

13. Arnold, A. P. (2009). The organizational-activational hypothesis as the foundation for a unified theory of sexual differentiation of all mammalian tissues. *Hormones and Behavior, 55,* 570–578.

14. Overall, XY women with androgen insensitivity syndrome are more likely to have longer-limb-to-trunk ratios, more typical of men. David Epstein reports conversations with endocrinologists in which they said that XY women with androgen insensitivity syndrome are also overrepresented in fashion modeling, due to their height and extra-long legs.

Epstein, D. (2013). *The sports gene: Inside the science of extraordinary athletic performance.* New York, NY: Current.

Don't let the lame title of this book scare you off. It's not a "genes explain everything" screed. It's a smart, clear, and thoughtful work.

15. One recent meta-analysis of testosterone levels claimed that the normal ranges of testosterone are 8.8 to 30.9 nanomoles per liter in males and 0.4 to 2.0 nanomoles per liter in females. However, this range encompasses not the entire population but that from the 2.5th to the 97.5th percentile. Obviously, individuals falling in the 97.5th to 100th percentiles will have higher levels, and this highest range is overrepresented in the pool of elite female athletes.

Clark, R. V., Wald, J. A., Swerdloff, R. S., Wang, C., Wu, F. C. W., Bowers, L. D., & Matsumoto, A. M. (2019). Large divergence in testosterone concentrations between men and women: Frame of reference for elite athletes in sex-specific competition in sports, a narrative review. *Clinical Endocrinology, 90,* 15–22.

16. Bermon, S., & Garnier, P. Y. (2017). Serum androgen levels and their relation to performance in track and field: Mass spectroscopy results from 2127 observations in male and female athletes. *British Journal of Sports Medicine, 51,* 1309–1314.

17. Handelsman, D. J. (2017). Sex differences in athletic performance emerge coinciding with the onset of male puberty. *Clinical Endocrinology, 87,* 68–72.

18. As of this writing in November 2019, the rules have changed yet again. In July 2019, the Court of Arbitration for Sport upheld the International Association of Athletics Federations' introduction of a new testosterone limit for female athletes who seek to compete in events ranging from four hundred meters to one mile in distance: five nanomoles per liter. To meet this standard would require Caster Semenya to take hormone-suppressing drugs, which she has refused to do. As a result, she was unable to defend her eight-hundred-meter title at the world championships in September 2019.

19. Sometimes, I like to imagine what nonhuman dating websites would look like:

Pathogenic Hottie. Hi everyone, I'm an *E. coli* bacterium of the strain O157:H7. I'm eight hours old and neither male nor female. I haven't posted a photo because I look just like everyone else. I have no religion and no Zodiac sign. I like eating glucose and keeping warm. Raw, undercooked, or spoiled meat is the best place to hang out! I'm not looking for a sexual relationship—when I want offspring I'll just split in two to clone myself. I'm looking for new friends to hang, divide frequently, and secrete Shiga toxin. And maybe watch *Game of Thrones*. Let's get together and give someone intestinal distress tonight!

20. There are actually several different modes of asexual reproduction in addition to binary fission (also called budding). Some animals can reproduce both sexually and asexually. Aphids, for example, reproduce sexually in most conditions, but when food is abundant in the spring then they switch to the more rapid asexual form called parthenogenesis, in which females lay eggs that can develop without fertilization to produce clones of the mother. This natural female cloning process is also found in other insect species, as well as some amphibians and fish.

21. Not only would asexual reproduction take away all those spousal arguments about who controls the TV remote, but once the TV was on there would be no option to choose a rom-com, due to a distinct lack of rom.

22. There are some genes for which this two-copy backup strategy doesn't work. One example includes genes on the X-chromosome, of which there is only one copy in males. Another example comes from those genes like the previously discussed *UBE3A*, in which only the maternal copy is expressed in certain cells (like neurons) so mutations in the maternal copy of the gene cannot be compensated for by the presence of a normal paternal copy. A related point is that when a bacterium with an important mutation divides and its progeny continue to divide and so on, that entire lineage will carry the mutation. The only way for that mutation to be eliminated from the population is for the entire lineage to die out. Perhaps that is not a problem for animals like bacteria and hydra that reproduce rapidly, but it's a major problem for an animal that requires months of gestation.

23. There are some animals (and many plants) that reproduce sexually, but where individuals produce both sperm and eggs. These are called hermaphrodites, and they include many species of worms and slugs and a few types of fish. Overall, it is estimated that of about 8.6 million animal species, about 65,000, or 0.7 percent, are hermaphroditic. To further complicate the situation, there are both simultaneous hermaphrodites, which produce eggs and sperm at the same time, and sequential hermaphrodites, which switch from producing sperm to eggs or vice versa.

24. Females can also compete with each other to attract the best mates, and this competition can drive the exaggeration of traits that signal fitness or fertility. These include body traits that emerge at sexual maturity, including fat deposits on the hips, breasts, and buttocks. As female fertility usually declines with age, these fertility signals can also sometimes mimic the characteristics of youth, like reduced body hair or a higher-pitched voice.

25. We all see the world through the lens of our experience, which is saturated with cultural ideas about men and women, boys and girls. Science seeks the objective truth, but it is performed by humans with biases, both conscious and subconscious, that influence our expectations, the types of questions we ask, and the design of our experiments. All scientists struggle to keep an open mind, and this struggle is ongoing. My point here is that, if some aspects of sexual selection theory—with promiscuous, aggressive, risk-taking men and sexually choosy, cooperative, nurturing women—turn out not to be true, this does not mean that the scientists who promote these theories are hateful, misogynistic knuckle draggers, it just means that they were wrong, as even the best scientists are on a regular basis.

26. Snyder, B. F., & Gowaty, P. A. (2007). A re-appraisal of Bateman's classic study of intrasexual selection. *Evolution, 61,* 2457–2468.

Gowaty, P. A., Kim, Y. K., & Anderson, W. W. (2012). No evidence of sexual selection in a repetition of Bateman's classic study of *Drosophila melanogaster. Proceedings of the National Academy of Science of the USA, 109,* 11740–11745.

Tang-Martínez, Z. (2016). Rethinking Bateman's principles: Challenging persistent myths of sexually reluctant females and promiscuous males. *Journal of Sex Research, 53,* 532–559.

27. This meta-analysis examined Bateman's three sexual selection metrics in seventy-two studies of sixty-six different species.

Janicke, T., Häderer, I. K., Lajeunese, M. J., & Anthes, N. (2016). Darwinian sex roles confirmed across the animal kingdom. *Science Advances, 2,* e1500983.

28. Jones, A. G., Rosenqvist, G., Berglund, A., Arnold, S. J., & Avise, J. C. (2000). The Bateman gradient and the cause of sexual selection in a sex-role-reversed pipefish. *Proceedings of the Royal Society of London, Series B, 267,* 677–680.

29. Emlen, S. T., & Wrege, P. H. (2004). Seize dimorphism, intrasexual competition, and sexual selection in wattled jacana (*Jacana jacana*) a sex-role -reversed shorebird in Panama. *The Auk, 121,* 391–403.

30. Other species where the males carry the eggs after fertilization include the Mormon cricket and the always-popular emperor penguin.

31. Clutton-Brock, T. (2009). Sexual selection in females. *Animal Behaviour, 77,* 3–11.

Tang-Martínez, Z. (2016). Rethinking Bateman's principles: Challenging persistent myths of sexually reluctant females and promiscuous males. *Journal of Sex Research, 53,* 532–559.

32. Hrdy, S. B. (1981). *The woman that never evolved.* Cambridge, MA: Harvard University Press.

33. The evolutionary biologist's term for presumably monogamous animals that engage multiple sexual partners is "sneaky fuckers." Really. It's even in the textbooks.

34. Larmuseau, M. H. D., Matthijs, K., & Wenseleers, T. (2016). Cuckolded fathers rare in human populations. *Trends in Ecology and Evolution, 31,* 327–329.

One might expect that the assignment of paternity would tend to be more accurate in groups where contraception is more readily available, but this does not appear to be the case. The authors write that cuckoldry rates "have stayed near constant at around 1% across several human societies over the past several hundred years."

35. Puts, D. (2016). Human sexual selection. *Current Opinion in Psychology, 7,* 28–32.

36. Brown, G. R., Laland, K. N., & Borgerhoff Mulder, M. (2009). Bateman's principles and human sex roles. *Trends in Ecology and Evolution, 24,* 297–304.

37. Polyandrous marriage (with one woman and multiple husbands) is rare. About 6 percent of societies have some form of polyandry some of the time, but those societies taken together account for less than 2 percent of worldwide population.

Starkweather, K., & Hames, R. (2012). A survey of non-classical polyandry. *Human Nature, 23,* 149–172.

38. Puts, D. (2016). Human sexual selection. *Current Opinion in Psychology, 7,* 28–32.

39. Jokela, M., Rotrirch, A., Rickard, I. J., Pettay, J., & Lummaa, V. (2010). Serial monogamy increases reproductive success in men but not in women. *Behavioral Ecology, 21,* 906–912.

40. Clark III, R. D., & Hatfield, E. (1989). Gender differences in receptivity to sexual offers. *Journal of Personality and Human Sexuality, 2,* 39–55.

Year later, after their 1989 paper had become widely cited, the authors looked back and told the story of the paper's origin, publishing travails, and ultimate impact. It's a fun read.

Clark III, R. D., & Hatfield, E. (2003). Love in the afternoon. *Psychological Inquiry, 14,* 227–231.

41. Too bad. Rainy weather is so romantic.

42. Hald, G. M., & Høgh-Olesen, H. (2010). Receptivity to sexual invitations from strangers of the opposite gender. *Evolution and Human Behavior, 31,* 453–458.

Guéguen, N. (2011). Effects of solicitor sex and attractiveness on receptivity to sexual offers: A field study. *Archives of Sexual Behavior, 40,* 915–919.

One replication in Austria involved somewhat older subjects (estimated ages in the midthirties) and only women were queried by men. In this study, 6 percent of the women approached agreed to sex with a male stranger.

Voracek, M., Hofhansl, A., & Fisher, M. L. (2005). Clark and Hatfield's evidence of women's low receptivity to male strangers' sexual offers revisited. *Psychological Reports, 97,* 11–20.

Years later, Elaine Hatfield and coworkers returned to this question using questions generated by a computer and accompanying blended faces. This is a very different design and so cannot truly be considered a replication. This time, 25 percent of men and 5 percent of women agreed to sex with a stranger. While the large difference between men and women remained, it's not clear if the reduction in the rate of male agreement was because of the times (2013 versus 1978), the use of computers versus real people, or some other factors.

Tappé, M., Bensman, L., Hayashi, K., & Hatfield, E. (2015). Gender differences in receptivity to sexual offers: A new research prototype. *Interpersona, 7.* doi:10.5964/ijpr.v7i2.121.

43. Would You . . . ? (Touch and Go song). (2019, December 23). Retrieved January 20, 2020, from Wikipedia: https://en.wikipedia.org/wiki/Would_You . . . %3F_(Touch_and_Go_song).

You can see the trippy video here: www.youtube.com/watch?v=izBbP2kro-c.

44. Fine, C. (2017). *Testosterone rex: Myths of sex, science, and society*. New York, NY: W. W. Norton.

Fine cites this study about the 11 percent incidence of female orgasm in heterosexual college-age hookups:

Armstrong, E. A., England, P., & Fogarty, A. C. (2012). Accounting for women's orgasm and sexual enjoyment in college hookups and relationships. *American Sociological Review, 77,* 435–462.

45. Herbenick, D., Reece, M., Schick, V., Sanders, S. A., Dodge, B., & Fortenberry, J. D. (2010). Sexual behavior in the United States: Results from a national probability sample of men and women ages 14–94. *The Journal of Sexual Medicine, 7,* 255–265.

46. Lyons, M., Lynch, A., Brewer, G., & Bruno, D. (2014). Detection of sexual orientation ("gaydar") by homosexual and heterosexual women. *Archives of Sexual Behavior, 43,* 345–352.

The result about lesbian versus straight women's casual sex behavior wasn't the main point of the Lyons et al. study but was one of their screening questions.

Bailey, J. M., Gaulin, S., Agyei, Y., & Gladue, B. A. (1994). Effects of gender and sexual orientation on evolutionarily relevant aspects of human mating psychology. *Journal of Personality and Social Psychology, 66,* 1081–1093.

47. For unusually clear, balanced, and nuanced reviews of this fraught topic, see:

Hines, M. (2010). Sex-related variation in human behavior and the brain. *Trends in Cognitive Sciences, 14,* 448–456.

Hines, M. (2020). Neuroscience and sex/gender. Looking back and forward. *Journal of Neuroscience, 40,* 37–43.

48. Hartog, J., Ferrer-i-Carbonell, A., & Jonker, N. (2002). Linking measured risk-aversion to individual characteristics. *Kyklos, 55,* 3–26.

49. Morgenroth, T., Fine, C., Ryan, M. K., & Genat, A. E. (2019). Sex, drugs, and reckless driving: Are measures biased toward identifying risk-taking in men? *Social Psychological and Personality Science, 9,* 744–753.

50. Hyde, J. S. (1984). How large are gender differences in aggression? A developmental meta-analysis. *Developmental Psychology, 20,* 722–736.

Archer, J. (2009). Does sexual selection explain human sex differences in aggression? *Behavioral and Brain Sciences, 32,* 249–311.

51. United Nations Office on Drugs and Crime. (2013). *Global study on homicide 2013*. Vienna, Austria: United Nations.

52. Wilson, M. L., et al. (2014). Lethal aggression in *Pan* is better explained by adaptive strategies than human impacts. *Nature, 513,* 414–417.

53. While there is no average difference between women and men in IQ test score, there is an interesting difference in the distribution: there are fewer men than women at the mean and slightly more men at the extremes, achieving the very worst and the very best scores. This result has been replicated several times in different populations and, at present, it remains a mystery with no convincing explanation for it. My guess is that it is caused by social factors: the very brightest

boys are encouraged and supported more than the very brightest girls and the very dullest boys are discouraged more than the very dullest girls. An alternative biological explanation is that men show more genetic variation than women generally. The idea is that, for mutations in genes on the X chromosome, women can balance out the effect in one copy with a normal second copy. Men, having a single X chromosome, do not have this safety net. It has been suggested that this is why, on average, men show greater variation in facial asymmetry than women. To be sure, this is all speculation at present—there is no clear link between genetic variation on the X chromosome and IQ test performance variance in men versus women.

Johnson, W., Carothers, A., & Deary, I. J. (2008). Sex differences in variability in general intelligence: A new look at the old question. *Perspectives on Psychological Science, 3,* 518–531.

54. Connellan, J., Baron-Cohen, S., Wheelwright, S., Bakti, A., & Ahluwalia, J. (2000). Sex differences in human neonatal social perception. *Infant Behavior and Development, 23,* 113–118.

55. One subsequent study found no sex difference in the time of face gaze in newborns (there was no object stimulus for comparison as in Connellan et al.). Interestingly, when a subset of these children was retested at thirteen to eighteen weeks of age, girls showed more eye contact time than boys. The interpretation of this latter finding is unclear. It could represent differential social learning in the early weeks of life (as the authors contend) or it could represent an innate difference between boys and girls that was not immediately expressed.

Leeb, R. T., & Rejskind, F. G. (2004). Here's looking at you, kid! A longitudinal study of perceived gender differences in mutual gaze behavior in young infants. *Sex Roles, 50,* 1–14.

56. Hines, M. (2009). Gonadal hormones and sexual differentiation of human brain and behavior. In D. W. Pfaff et al. (Eds.), *Hormones, brain and behavior* (2nd ed.) (pp. 1869–1909). Cambridge, MA: Academic Press.

57. Jordan-Young, R. M. (2010). *Brainstorm: The flaws in the science of sex differences* (pp. 246–255). Cambridge, MA: Harvard University Press.

58. Pasterski, V. L., Geffner, M. E., Brain, C., Hindmarsh, P., Brook, C., & Hines, M. (2005). Prenatal hormones and postnatal socialization by parents as determinants of male-typical toy play in girls with congenital adrenal hyperplasia. *Child Development, 76,* 264–278.

59. Hines, M. (2010). Sex-related variation in human behavior and the brain. *Trends in Cognitive Sciences, 14,* 448–456.

60. Alexander, G. M., & Hines, M. (2002). Sex differences in response to children's toys in nonhuman primates (*Cercopthecus aethiops sabaeus*). *Evolution and Human Behavior, 23,* 467–479.

61. Arnold, A. P., & McCarthy, M. M. (2016). Sexual differentiation of the brain and behavior: A primer. In D. W. Pfaff & N. D. Volkow (Eds.), *Neuroscience in the 21st century* (pp. 2139–2168). New York, NY: Springer.

In another twist on the potential androgen-socialization interaction, Melissa Hines and her coworkers showed that girls exposed to high concentrations of prenatal androgens as a result of congenital adrenal hyperplasia showed reduced imitation of female models choosing particular objects:

Hines, M., et al. (2016). Prenatal androgen exposure alters girls' responses to information indicating gender-appropriate behavior. *Philosophical Transactions of the Royal Society of London, Series B, 371,* 20150125.

62. Arnold, A. P., & McCarthy, M. M. (2016). Sexual differentiation of the brain and behavior: A primer. In D. W. Pfaff & N. D. Volkow (Eds.), *Neuroscience in the 21st century* (pp. 2139–2168). New York, NY: Springer.

63. Neurological disorders like stuttering, Tourette's, and dyslexia occur two to three times more often in boys, and yet, unlike Parkinson's, have no known link to either hormonal or environmental influences. They are also not so subject to diagnostic bias as ADD and ADHD can be.

64. Heflin, C. M., & Iceland, J. (2009). Poverty, material hardship and depression. *Social Science Quarterly, 90,* 1051–1071.

65. Baron-Cohen, S., Lutchmaya, S., & Knickmeyer, R. C. (2004). *Prenatal testosterone in mind: Amniotic fluid studies.* Cambridge, MA: MIT Press.

Auyeung, B., Ahluwalia, J., Thomson, L., Taylor, K., Hackett, G., O'Donnell, K. J., & Baron-Cohen, S. (2012). Prenatal versus postnatal sex steroid hormone effects on autistic traits in children at 18 to 24 months of age. *Molecular Autism, 3,* 17.

66. Kung, K. T., et al. (2016). No relationship between prenatal androgen exposure and autistic traits: Convergent evidence from studies of children with congenital adrenal hyperplasia and amniotic testosterone concentrations in typically developing children. *Journal of Child Psychiatry and Psychology, 57,* 1455–1462.

67. Rodeck, C. H., Gill, D., Rosenberg, D. A., & Collins, W. P. (1985). Testosterone levels in midtrimester maternal and fetal plasma and amniotic fluid. *Prenatal Diagnosis, 5,* 175–181.

68. There are a few cases where invasive studies of the human brain are possible, but only rarely and only in the context of disease. For example, some patients have consented to have brief recordings made from electrodes inserted in their brains during the course of neurosurgery. Some conditions, like intractable epilepsy, necessitate the removal of brain tissue, which can then be kept alive for a few hours and studied with electrodes or other techniques.

69. Shansky, R. M., & Woolley, C. S. (2016). Considering sex as a biological variable will be valuable for neuroscience research. *Journal of Neuroscience, 36,* 11817–11822.

70. Arnold, A. P., & McCarthy, M. M. (2016). Sexual differentiation of the brain and behavior: A primer. In D. W. Pfaff & N. D. Volkow (Eds.), *Neuroscience in the 21st century* (pp. 2139–2168). New York, NY: Springer.

Hines, M. (2009). Gonadal hormones and sexual differentiation of human brain and behavior. In D. W. Pfaff et al. (Eds.), *Hormones, brain and behavior* (2nd ed.) (pp. 1869–1909). Cambridge, MA: Academic Press.

71. Lotze, M., Domin, M., Gerlach, F. H., Gaser, C., Lueders, E., Schmidt, C. O., & Neumann, N. (2018). Novel findings from 2,838 adult brains on sex differences in gray matter brain volume. *Scientific Reports, 9,* 1671.

72. Wheelock, M. D., Hect, J. L., Hernandez-Andrade, E., Hassan, S. S., Romero, R., Eggebrecht, A. T., & Thomason, M. E. (2019). Sex differences in functional connectivity during fetal brain development. *Developmental Cognitive Neuroscience, 36,* 100632.

73. Joel, D., et al. (2015). Sex beyond the genitalia: The human brain mosaic. *Proceedings of the National Academy of Sciences of the USA, 112,* 15468–15473.

74. Chekroud, A. M., Ward, E. J., Rosenberg, M. D., & Holmes, A. J. (2016). Patterns in the human brain mosaic discriminate males from females. *Proceedings of the National Academy of Sciences of the USA, 113,* e1968.

75. Mitchell, K. J. (2018). *Innate: How the wiring of our brains shapes who we are* (pp. 196–198). Princeton, NJ: Princeton University Press.

76. This is most easily seen in cultures that have enshrined nonbinary gender status, such as Native American two-spirit and Polynesian *mahu.*

77. Flores, A. R., Herman, J. L., Gates, G. J., & Brown, T. N. T. (2016). *How many adults identify as transgender in the United States?* Los Angeles, CA: The Williams Institute.

This analysis may also reflect the limits of self-reporting in obtaining accurate data, even with an anonymous survey. In some states seen as more accepting, more adults identified themselves as transgender, and in some states perceived as more resistant, fewer adults did so. The percentage of adults identifying as transgender by state ranged from 0.30 percent in North Dakota to 0.78 percent in Hawaii. Importantly, the youngest age group surveyed, eighteen-to-twenty-four-year-olds, were most likely to identify as transgender, suggesting a societal shift. At present, we don't have a reliable measurement of the number of transgender teenagers or children.

78. Cross-dressing may be an expression of gender dysphoria or it may be a more subtle expression of individual identity—you don't necessarily have to feel gender dysphoric to cross-dress. You might just enjoy the artistry of it, or enjoy tweaking people's social expectations.

79. Ben transitioned in 1998 and, I'm happy to say, continued his brilliant career in neuroscience with broad acceptance from his friends and colleagues until his death in 2017. His story is told in:

Barres, B. (2018). *The autobiography of a transgender scientist.* Cambridge, MA: MIT Press.

80. Imperato-McGinley, J., Peterson, R. E., Gautier, T., & Sturla, E. (1979). Androgens and the evolution of male-gender identity among male pseudohermaphrodites with 5-alpha-reductase deficiency. *The New England Journal of Medicine, 300,* 1233–1237.

Several clusters of 5-alpha reductase deficiency have been found around the world, in the Taurus Mountains of southern Turkey, in the southwestern portion of the Dominican Republic, and among the Simbari Anga people in the eastern highlands of Papua New Guinea.

81. Brocca, M. E., & Garcia-Segura, L. M. (2018). Non-reproductive functions of aromatase in the central nervous system under physiological and pathological conditions. *Cellular and Molecular Neurobiology.* doi:10.1007/s10571-018-0607-4.

Aromatase is the enzyme that converts testosterone into estradiol (a potent form of estrogen). Its presence in the brain means that brain estrogen signaling is not entirely eliminated in females, even when the ovaries are surgically removed. It also means that estrogen signaling is present in male brains as well, although its effects are not always the same.

82. Coolidge, F. L., Thede, L. L., & Young, S. E. (2002). The heritability of gender identity disorder in a child and adolescent twin sample. *Behavioral Genetics, 32,* 251–257.

Heylens, G., et al. (2012). Gender identity disorder in twins: A review of the case report literature. *Journal of Sexual Medicine, 9,* 751–757.

Gómez-Gil, E., Esteva, I., Almaraz, M. C., Pasaro, E., Segovia, S., & Guillamon, A. (2010). Familiality of gender identity disorder in twins. *Archives of Sexual Behavior, 39,* 546–552.

83. Zhou, J. N., Hofman, M. A., Gooren, L. J., & Swaab, D. F. (1995). A sex difference in the human brain and its relation to transsexuality. *Nature, 378,* 68–70.

A particular subdivision of the BNST is sexually dimorphic: the central complex, known as BNSTc.

84. Chung, W. C., De Vries, G. J., & Swaab, D. F. (2002). Sexual differentiation of the bed nucleus of the stria terminalis in humans may extend into adulthood. *Journal of Neuroscience, 22,* 1027–1033.

85. Smith, E. S., Junger, J., Derntl, B., & Habel, U. (2015). The transsexual brain—a review of findings on the neural basis of transsexualism. *Neuroscience and Biobehavioral Reviews, 59,* 251–266.

Chapter Five. Who Do You Love?

1. The name has been changed to protect the guilty. This joke is showing its age (circa 1978), as we now recognize a spectrum of gender identity. Perhaps, in present lingo, Jane would best be described as both "pansexual" and "sapiosexual," that is, attracted to intelligence. As she explained it to me: "If I meet someone and they say something clever then I'll probably want to jump their bones."

2. Herdt, G. H. (1984). *Ritualized homosexuality in Melanesia.* Berkeley, CA: University of California Press.

There is evidence that, more recently, ritual, intergenerational male homosexuality in Melanesia has largely ceased among those groups with extensive Western and Christian contact, like the Gebusi people of Papua New Guinea.

Knauft, B. M. (2003). What ever happened to ritualized homosexuality? Modern sexual subjects in Melanesia and elsewhere. *Annual Review of Sex Research, 14,* 137–159.

Of course, ritualized intergenerational male homosexuality is not just found in Melanesia. There are also examples from traditional cultures in Australia and the Amazon basin.

Unfortunately, we know much less about ritual female homosexuality. Of course, ideas about sexual orientation and intergenerational male homosexuality have changed in many cultures through time, with notable examples from ancient Greece and medieval Japan.

3. Van Anders, S. M. (2015). Beyond sexual orientation: Integrating gender/sex and diverse sexualities via sexual configurations theory. *Archives of Sexual Behavior, 44,* 1177–1213.

4. Laumann, E. O., Gagnon, J. H., Michael, R. T., & Michaels, S. (1994). *The social organization of sexuality: Sexual practices in the United States.* Chicago, IL: University of Chicago Press.

Interestingly, more recent random anonymous surveys, conducted in times when societal approval of same-sex attraction has generally increased, have not indicated a higher incidence of homosexuality and bisexuality.

5. Lever, J. (1994, August 23). Sexual revelations: The 1994 *Advocate* survey of sexuality and relationships: The men. *The Advocate*, 17–24.

Neuroscientist Simon LeVay has stated the matter succinctly: "If their sexual orientation was indeed a choice, gay people should remember having made it. But, by and large, they don't."

LeVay, S. (2010). *Gay, straight, and the reason why: The science of sexual orientation* (p. 41). Oxford, UK: Oxford University Press.

For a nice explanation by some who state that they have indeed chosen their homosexual orientation, see www.queerbychoice.com/.

6. Cole, D. (2019, April 8). Buttigieg to Pence: "If you got a problem with who I am, your problem is not with me—your quarrel, sir, is with my creator." CNN. Retrieved from www.cnn.com/2019/04/08/politics/pete-buttigieg-mike-pence/index.html.

7. The quotations from the *Obergefell v. Hodges* 2015 decision are taken from:

Diamond, L. M., & Rosky, C. J. (2016). Scrutinizing immutability: Research on sexual orientation and U.S. legal advocacy for sexual minorities. *Journal of Sex Research, 53,* 363–391.

The Diamond and Rosky article does a great job of summarizing the state of knowledge about the immutability of sexual orientation and also arguing that immutability should not be the basis for supporting civil rights for gay and bi people.

8. Diamond, L. M. (2008). Female bisexuality from adolescence to adulthood: Results from a 10 year longitudinal study. *Developmental Psychology, 44,* 5–14.

Diamond, L. M. (2008). *Sexual fluidity: Understanding women's love and desire.* Cambridge, MA: Harvard University Press.

9. Does the existence of sexual fluidity mean that, at least for women, sexual orientation could be changed by so-called reparative therapy, as is sometimes promoted by fundamentalist religious groups? The answer appears to be no. When a task force of the American Psychological Association studied the scientific literature, their conclusion was that "enduring change to an individual's sexual orientation was unlikely" following such treatment in either women or men.

APA Task Force on Appropriate Therapeutic Responses to Sexual Orientation. (2009). *Report of the Task Force on Appropriate Therapeutic Responses to Sexual Orientation.* Washington, DC: APA Press.

To be clear, this doesn't mean that people can't change their behavior to conform with religious or cultural ideals. Rather, it means that one cannot manufacture opposite-sex attraction out of whole cloth or diminish same-sex attraction through training. In much the same way, Roman Catholic priests can often be celibate even though they experience sexual urges.

10. Rosenthal, A. M., Sylva, D., Safron, A., & Bailey, J. M. (2012). The male bisexuality debate revisited: Some bisexual men have bisexual arousal patterns. *Archives of Sexual Behavior, 41,* 135–147.

This report follows others from the same group, which found that most bisexual men have either a gay or a straight arousal pattern but not a truly bisexual one. It is likely that the difference results from the more stringent inclusion criteria in this 2012 study: "Bisexual participants were required to have had at least two sexual partners of each sex and a romantic relationship of at least 3 months duration with at least one member of each sex."

11. Bailey, J. M. (2009). What is sexual orientation and do women have one? In D. A. Hope (Ed.) *Contemporary perspectives on lesbian, gay and bisexual identities.* New York, NY: Springer.

12. Suschinsky, K. D., Dawson, S. J., & Chivers, M. J. (2017). Assessing the relationship between sexual concordance, sexual attractions and sexual identity in women. *Archives of Sexual Behavior, 46,* 179–192.

This is not to say that all lesbian women report a lack of arousal to gay male or hetero porn. In fact, there seems to be a particular appreciation of the former among some lesbians.

Neville, L. (2015). Male gays in the female gaze: Women who watch m/m pornography. *Porn Studies, 2,* 192–207.

13. Bouchard, K. N., Chivers, M. L., & Pukall, C.F. (2017). Effects of genital response measurement device and stimulus characteristics on sexual concordance in women. *The Journal of Sex Research, 54,* 1197–1208.

14. I, for one, find those bonobo sex videos to be hot.

15. Chivers, M. L. (2017). The specificity of women's sexual response and its relationship with sexual orientation: A review and ten hypotheses. *Archives of Sexual Behavior, 46,* 1161–1179.

In addition to the "preparation hypothesis" to explain women's discordant vaginal responses, Chivers suggests nine others, all of which are interesting and useful.

16. This section, on the determinants of sexual orientation, is adapted from Linden, D. J. (2018). Human sexual orientation is strongly influenced by biological factors. In D.J. Linden, (Ed.) *Think tank: Forty neuroscientists explore the biological roots of human experience* (pp. 215–224). New Haven, CT: Yale University Press.

17. Tasker, F. L., & Golombok, S. (1997). *Growing up in a lesbian family: Effects on child development.* New York, NY: Guilford Press.

Green, R., Mandel, J. B., Hotvedt, M. E., Gray, J., & Smith, L. (1986). Lesbian mothers and their children: A comparison with solo parent heterosexual mothers and their children. *Archives of Sexual Behavior, 7,* 175–181.

18. Patterson, C. J. (2005). *Lesbian and gay parents and their children: Summary of research findings.* Washington, DC: American Psychological Association.

19. Brannock, J. C., & Chapman, B. E. (1990). Negative sexual experiences with men among heterosexual women and lesbians. *Journal of Homosexuality, 19,* 105–110.

Stoddard, J. P., Dibble, S. L., & Fineman, N. (2009). Sexual and physical abuse: A comparison between lesbians and their heterosexual sisters. *Journal of Homosexuality, 56,* 407–420.

20. Isay, R. A. (1999). Gender development in homosexual boys: Some developmental and clinical considerations. *Psychiatry, 62,* 187–194.

21. Bailey, J. M., & Pillard, R.C. (1991). A genetic study of male sexual orientation. *Archives of General Psychiatry, 48,* 1089–1096.

Bailey, J. M., & Benishay, D. S. (1993). Familial aggregation of female sexual orientation. *American Journal of Psychiatry, 150,* 272–277.

22. Långström, N., Rahman, Q., Carlström, E., & Lichtenstein, P. (2010). Genetic and environmental effects on same-sex sexual behavior: A population study of twins in Sweden. *Archives of Sexual Behavior, 39,* 75–80.

This study is notable because of its large sample size and random sampling of twins.

Another study, which used a large randomized population of British female twins, yielded a similar estimate for the degree of heritability of sexual orientation in women: 25 percent.

Burri, A., Cherkas, L., Spector, T., & Rahman, Q. (2011). Genetic and environmental influences on female sexual orientation, childhood gender typicality and adult gender identity. *PLOS One, 6,* e21982.

23. Recently, a large GWAS was conducted on a population of men and women from Europe and the United States to identify gene variations that might contribute to same-sex attraction. The authors found gene variants accounting for about 1 percent of variation in this trait. However, in my view, this study is flawed because it used the question "Have you *ever* had sex with someone of the same sex as yourself?" as the basis for sorting subjects (italics mine). This is a much larger and more variable population of people than those who identify as consistently gay or bi.

Ganna, A., et al. (2019). Large-scale GWAS reveals insights into the genetic architecture of same-sex sexual behavior. *Science, 365,* eaat7693.

24. Blanchard, R. (2018). Fraternal birth order, family size, and male homosexuality: Meta-analysis of studies spanning 25 years. *Archives of Sexual Behavior, 47,* 1–15.

Having multiple older brothers didn't further increase the chances of same-sex attraction beyond that conferred by a single older brother, but it did increase the chance of other gender-atypical behavior.

25. Bogaert, A. F., et al. (2018). Male homosexuality and maternal immune responsivity to the Y-linked protein NLGN4Y. *Proceedings of the National Academy of Sciences of the USA, 115,* 302–306.

26. Meyer-Bahlburg, H. F., Dolezal, C., Baker, S. W., & New, M. I. (2008). Sexual orientation in women with classical or non-classical congenital adrenal hyperplasia as a function of degree of prenatal androgen excess. *Archives of Sexual Behavior, 37,* 85–99.

27. Allen, L. S., & Gorski, R. A. (1992). Sexual orientation and the size of the anterior commissure in the human brain. *Proceedings of the National Academy of Sciences of the USA, 89,* 7199–7202.

LeVay, S. (1991). A difference in hypothalamic structure between heterosexual and homosexual men. *Science, 253,* 1034–1037.

28. Byne, W., Tobet, S., Mattiace, L. A., Lasco, M. S., Kemether, E., Edgar, M. A., Morgello, S., Buchsbaum, M. S., & Jones, L. B. (2001). The interstitial nuclei of the human anterior hypothalamus: An investigation of variation with sex, sexual orientation, and HIV status. *Hormones and Behavior, 40,* 86–92.

Lasco, M. S., Jordan, T. J., Edgar, M. A., Petito, C. K., & Byne, W. (2002). A lack of dimorphism of sex or sexual orientation in the human anterior commissure. *Brain Research, 936,* 95–98.

29. Not to belabor a point here but, of course, any differences measured in adults could be innate, result from life experience, or result from interaction between factors present at birth and life experience.

30. Connellan, J., Baron-Cohen, S., Wheelwright, S., Batki, A., & Ahluwalia, J. (2001). Sex differences in human neonatal social perception. *Infant Behavior and Development, 23,* 113–118.

31. Green, R. (1987). *The "sissy-boy syndrome" and the development of homosexuality.* New Haven, CT: Yale University Press.

Drummond, K. D., Bradley, S. J., Peterson-Badali, M., & Zucker, K. J. (2008). A follow-up study of girls with gender identity disorder. *Developmental Psychology, 44,* 34–45.

32. Karlson, O., & Lüscher, M. (1959). "Pheromones": A new term for a class of biologically active substances. *Nature, 183,* 55–56.

33. The isolation and characterization of bombykol was a truly heroic effort that took about twenty years and used over five hundred thousand silkworm moths.

34. Wyatt, T. D. (2014). *Pheromones and animal behavior: Chemical signals and signatures* (2nd ed.). Cambridge, UK: Cambridge University Press.

35. The overarching term for chemical signals between organisms is "semiochemical." Signals between members of the same species are either pheromones, if they produce an innate stereotyped response (either a behavior or a developmental process), or signature mixtures that must be learned by the receiver to recognize an individual or member of a social group like a family or colony. Signals between members of different species are called "allelochemicals."

Wyatt, T. D. (2014). *Pheromones and animal behavior: Chemical signals and signatures* (2nd ed.). Cambridge, UK: Cambridge University Press.

36. There are also pheromones that don't act through the chemical senses (nose, antennae) at all, but rather are transferred by feeding (as in the case of royal jelly in honeybees) or by receiving semen during mating (as in some snakes and flies).

37. Wyatt, T. D. (2014). *Pheromones and animal behavior: Chemical signals and signatures* (2nd ed.). Cambridge, UK: Cambridge University Press.

38. McClintock, M. K. (1971). Menstrual synchrony and suppression. *Nature, 229,* 244–245.

39. Stern, K., & McClintock, M. K. (1998). Regulation of ovulation by human pheromones. *Nature, 392,* 177–179.

40. Wyatt, T. D. (2015). The search for human pheromones: The lost decades and the necessity of returning to first principles. *Proceedings of the Royal Society, Series B, 282,* 20142994.

The money quote from this paper: "If we are to find pheromones we need to treat ourselves just as if we were a newly discovered mammal."

41. Roberts, S. A., Simpson, D. M., Armstrong, S. D., Davidson, A. J., McLean, L., Beynon, R. J., & Hurt, J. L. (2010). Darcin: A male pheromone that stimulates female memory and sexual attraction to an individual male's odor. *BMC Biology, 8,* 75.

42. Doucet, S., Soussignan, R., Sagot, P., & Schall, B. (2009). The secretion of areolar (Montgomery's) glands from lactating women elicits selective unconditional responses in neonates. *PLoS One, 4,* e7579.

43. Charra, R., Datiche, F., Casthano, A., Gigot, V., Schall, B., & Coureaud, G. (2012). Brain processing of the mammary pheromone in newborn rabbits. *Behavioral Brain Research, 226,* 1790188.

Matsuo, T., Hattori, T., Asaba, A., Inoue, N., Kanomata, N., Kikusui, T., Kobayakawa, R., & Kobayakawa, K. (2015). Genetic dissection of pheromone processing reveals main olfactory system-mediated social behaviors in mice. *Proceedings of the National Academy of Sciences of the USA, 112,* E311–E320.

44. Wyatt, T. D. (2015). The search for human pheromones: The lost decades and the necessity of returning to first principles. *Proceedings of the Royal Society, Series B, 282,* 20142994.

45. Gilbert, A. N., Yamazaki, K., Beauchamp, G. K., & Thomas, L. (1986). Olfactory discrimination of mouse strains (*Mus musculus*) and major histocompatibility types by humans (*Homo sapiens*). *Journal of Comparative Psychology, 100,* 262–265.

46. Gilbert, A. (2014). *What the nose knows: The science of scent in everyday life.* Fort Collins, CO: Synesthetics, Inc.

47. Wedekind, C., Seebeck, T., Bettens, F., & Paepke, A. J. (1995). MHC-dependent mate preferences in humans. *Proceedings of the Royal Society of London, Series B, 260,* 245–249.

Interestingly, this preference was reversed in women taking oral hormonal contraceptives.

48. Wedekind, C., & Furi, S. (1997). Body odour preferences in men and women: Do they aim for specific MHC combinations or simply hetrozygosity? *Proceedings of the Royal Society of London, Series B, 264,* 1471–1479.

49. Probst, F., Fischbacher, U., Lobmaier, J. S., Wirthmüller, U., & Knoch, D. (2017). Men's preference for women's body odours are not associated with human leukocyte antigen. *Proceedings of the Royal Society of London, Series B, 284,* 20171830.

Wedekind, C. (2018). A predicted interaction between odour pleasantness and intensity provides evidence for major histocompatibility complex social signalling in women. *Proceedings of the Royal Society of London, Series B, 285,* 20172714.

Lobmaier, J. S., Fischbacher, U., Probst, F., Wirthmuller, U., & Knoch, D. (2018). Accumulating evidence suggests that men do not find body odours of human leukocyte dissimilar women more attractive. *Proceedings of the Royal Society of London, Series B, 285,* 21080566.

50. Cretu-Stancu, M., Kloosterman, W. P., & Pulit, S. L. (2018). No evidence that mate choice in humans is dependent on the MHC. *BioArXiv*. Advance online publication. Retrieved from www.biorxiv.org/node/103128.abstract.

51. Gangestad, S. W., & Buss, D. M. (1993). Pathogen prevalence and human mate preferences. *Ethology and Sociobiology, 14,* 89–96.

52. Winkelmann, R. K. (1959). The erogenous zones: Their nerve supply and significance. *Proceedings of the Staff Meetings of the Mayo Clinic, 34,* 39–47.

Krantz, K. E. (1958). Innervation of the human vulva and vagina: A microscopic study. *Obstetrics and Gynecology, 12,* 382–396.

Martin-Alguacil, N., Pfaff, D. W., Shelly, D. N., & Schober, J. M. (2008). Clitoral sexual arousal: An immunocytochemical and innervation study of the clitoris. *BJU International, 191,* 1407–1413.

Halata, Z., & Munger, B. L. (1986). The neuroanatomical basis for protopathic sensibility of the human glans penis. *Brain Research, 371,* 205–230.

Chapter Six. We Are the Anti-Pandas

1. This made me curious: Was it just a woman thing? I made a quick, minimal, woman-seeking-man profile so I could look at straight guy postings. They, too, tended to list their favorite and least favorite foods, sometimes at length. Upon further inquiry, I learned that men and women, gay and straight and bi, were all talking about their individual food preferences when searching for a date.

2. This usage is not just a quirk of English. It appears in many, but not all, languages across the world.

3. "Notably, patients who complained only of loss of the sense of taste and who had demonstrable olfactory or gustatory loss were nearly three times more likely to show an olfactory deficit than a gustatory one."

Dees, D. A., et al. (1991). Smell and taste disorders, a study pf 750 patients from the University of Pennsylvania smell and taste center. *Archives of Otolaryngolgy, Head and Neck Surgery, 117,* 5190528.

4. However, not even the smartest sea anemone can accompany the disgust face with the Yiddish expression "*Feh!*" in the style of my grandmother.

5. In addition to the five basic tastes—sweet, sour, bitter, salt, umami—for which sensors have been identified, there are suggestions that there may be oral sensors for fat, calcium, and carbohydrates as well. However, at present, this is more supposition than fact. Genes that encode functional oral carbohydrate, calcium, or fat sensors have yet to be clearly identified.

6. If you're like me, your high school science textbook had a drawing showing the tongue divided up into different taste zones: bitter at the back, sweet in the front, and salty and sour on the sides. Not to put too fine a point on it, but this is nonsense. There are some subtle differences in taste sensation across the surface of the tongue, but nothing like what was depicted. You can do an experiment at home to convince yourself. Dab a cotton swab in sour vinegar or sweet sugar crystals and then touch it to various parts of your tongue. You'll find that there is little variation in the taste sensation across the tongue surface. If you are interested in how this mistaken tongue map came to be so widely known, you can read Steven Munger's nice article on the topic:

Munger, S. (2017, May 23). The taste map of the tongue you learned in school is all wrong. *Smithsonian Magazine*. Retrieved from www.smithsonianmag.com/science-nature/neat-and-tidy-map-tastes-tongue-you-learned-school-all-wrong-180963407/.

The assertion that the electrical signals produced when the individual taste cells are activated by food or drink are kept separate as they are conveyed to the brain is certainly true in most cases, but some argument remains about whether it is true in all cases.

7. You might be wondering why I'm using the term "taste sensor" rather than "taste receptor." The reason is that while the sensors for bitter, sweet, and umami taste are indeed receptors (they bind the relevant chemicals on the outside of the cell and then change their shape to pass a signal across the membrane), the sensors for sour and salt are not. They are ion channels that allow the H^+ and Na^+ ions, respectively, to pass into the taste receptor cell rather than keeping them outside. To make life even more complicated, the sweet, bitter, and umami receptors are dimers, meaning that they are formed by the combination of two receptor proteins, either of the same type (homodimers) or different types (heterodimers). The sweet receptors can either be a TAS1R2-TAS1R3 heterodimer or a TAS1R3 homodimer, the umami receptor is a TAS1R1-TAS1R3 heterodimer, and the bitter receptors are formed as hetero- or homodimers of the various products of at least twenty-five different genes of the *TAS2R* family.

8. Bitter taste is an indicator of some but not all bacterial food infections. In fact, the most common lethal bacteria in foods—including salmonella, listeria, staphylococcus, and shigella—are tasteless and odorless.

Breslin, P. A. S. (2019). *Chemical senses in feeding, belonging and surviving: Or, are you going to eat that?*. Cambridge, UK: Cambridge University Press.

Strong bitter tastes can produce sensations of nausea, which are highly aversive.

Peyrot de Gachons, C., Beauchamp, G. K., Stern, R. M., Koch, K. L., & Breslin, P. A. S. (2011). Bitter taste induces nausea. *Current Biology, 21,* R247–R248.

9. Chandrashekar, J., Kuhn, C., Oka, Y., Yarmolinsky, D. A., Hummler, E., Ryba, N. J. P., & Zuker, C. S. (2010). The cells and peripheral representation of sodium taste in mice. *Nature, 464,* 297–301.

There is still some debate about whether ENaC is also the low-sodium taste sensor in humans.

10. At the time of this writing, only one sour taste sensor has been clearly identified, a proton-passing ion channel called OTOP1.

Tu, Y. H., et al. (2018). An evolutionarily conserved gene family encodes proton-selective ion channels. *Science, 359,* 1047–1050.

Others have claimed that additional membrane proteins also serve this function, and the argument has yet to be settled. It's possible that additional sour sensors will be identified (there are already two other members of the OTOP family that are known). Perhaps, like salty taste, different sensors will be found for mildly sour and strongly sour solutions. But there are very unlikely to be anywhere near twenty-five of them.

11. If we step back a bit, we can realize that the taste sensor proteins are actually just detectors of chemicals. When they are present in the taste receptor cells of the tongue, which are wired to the gustatory centers of the brain, they subserve taste sensations. But when they are deployed elsewhere, they can have other important functions. For example, bitter sensors, like the aforementioned T2R38, can detect chemicals released by clusters of bacteria and are present in the gums, lungs, trachea, sinuses, and the gut. When activated in the respiratory pathways, they trigger innate immune responses and relax the airway to allow for more vigorous coughing to expel bacteria. There are also taste sensors in the gut, the skin, and several other locations.

Lu, P., Zhang, C. H., Lifshitz, L. M., & ZhuGe, R. (2017). Extraoral bitter taste receptors in health and disease. *Journal of General Physiology, 149,* 181–197.

In addition, there are suggestions that people with a malfunctioning *T2R38* gene are more susceptible to serious, chronic sinus infections requiring surgery.

Adappa, N. D., et al. (2014). The bitter taste receptor T2R38 is an independent risk factor for chronic rhinosinusitis requiring sinus surgery. *International Forum of Allergy & Rhinology, 4,* 3–7.

12. Feng, P., Zheng, J., Rossiter, S. J., Wang, D., & Zhao, H. (2014). Massive losses of taste receptor genes in toothed and baleen whales. *Genome Biology and Evolution, 6,* 1254–1265.

Seven species of toothed whale and five species of baleen whale were examined and all had lost sour, bitter, sweet, and umami tastes. The *ENaC* gene was intact, but this gene also functions to maintain sodium balance in the kidney, so it's still unclear if it is expressed in the oral cavity of whales and subserves salty taste.

13. For you hard-core nerds out there, the processing stations of the taste pathway between the taste ganglia and the insular cortex are the nucleus of the solitary tract, the parabrachial nucleus, and the ventral posterior thalamus.

14. Wang, L., et al. (2018). The coding of valence and identity in the mammalian taste system. *Nature, 558,* 127–131.

15. Lee, H., Macpherson, L. J., Parada, C. A., Zuker, C. S., & Ryba, N. J. P. (2017). Rewiring the taste system. *Nature, 548,* 330–333.

16. Lugaz, O., Pillias, A. M., & Faurion, A. (2002). A new specific ageusia: Some humans cannot taste L-glutamate. *Chemical Senses, 27,* 105–115.

17. Shigemura, N., Shirosaki, S., Sanematsu, K., Yoshida, R., & Ninomiya, Y. (2009). Genetic and molecular basis of individual differences in human umami taste perception. *PLoS One, 4,* e6717.

Raliou, M., et al. (2011). Human genetic polymorphisms in T1R1 and T1R3 taste receptor subunits affect their function. *Chemical Senses, 36,* 527–537.

18. Fushan, A. A., Simons, C. T., Slack, J. P., Manichaikul, A., & Drayna, D. (2009). Allelic polymorphism within the *TAS1R3* promoter is associated with human taste sensitivity to sucrose. *Current Biology, 19,* 1288–1293.

This is a slightly different situation than the genetic variation in the umami receptor. In that case, the variation caused changes in the structure of the sensor protein itself. Here, the variation is in the promoter region of the gene, which means that it will impact not the structure of the sweet sensor but the amount of it that is produced in taste receptor cells.

19. Roura, E., et al. (2015). Variability in human bitter taste sensitivity to chemically diverse compounds can be accounted for by differential TAS2R activation. *Chemical Senses, 40,* 427–435.

20. PROP is an abbreviation for 6-n-propylthiouracil. It is a drug used to treat hyperthyroidism. When used in taste experiments, the dose used is much smaller than that employed clinically.

21. The genetics underlying the density of fungiform papillae remains unclear. There is a statistical association of high-density supertasting with certain variants of the *T2R38* gene, but this only explains a portion of the variance. Also, there is no molecular or cellular explanation for why *T2R38* variation could cause development of more fungiform papillae, so the two measures may be statistically but not causally related.

Hayes, J. E., Bartoshuk, L. M., Kidd, J. R., & Duffy, V. B. (2008). Supertasting and PROP bitterness depends on more than the *TASR38* gene. *Chemical Senses, 33,* 255–265.

There is another wrinkle. While the density of fungiform papillae is positively correlated with individual sensitivity to PROP, it does not predict individual sensitivity to quinine, another bitter chemical.

Delwiche, J. F., Buletic, Z., & Breslin, P. A. S. (2001). Relationship of papillae number to bitter intensity of quinine and PROP within and between individuals. *Physiology & Behavior, 74,* 329–337.

22. Bartoshuk, L. M., Cunningham, K. E., Dabrila, G. M., Duffy, V. B., Etter, L., Fast, K. R., Lucchina, L. A., Prutkin, J. M, & Snyder, D. J. (1999). From sweets to hot peppers: Genetic variation in taste, oral pain, and oral touch. In G. A. Bell & A. J. Watson (Eds.), *Tastes and aromas* (pp. 12–22). Sydney, Australia: University of New South Wales Press.

23. De Graaf, C., & Zandstra, E. (1999). Sweetness intensity and pleasantness in children, adolescents, and adults. *Physiology & Behavior, 67,* 513–520.

24. Prutkin, J., et al. (2000). Genetic variation and inferences about perceived taste intensity in mice and men. *Physiology & Behavior, 69,* 161–173.

While the evidence for increased bitter taste sensitivity in women during the first trimester of pregnancy is solid, claims that bitter sensitivity varies systematically over the ovarian cycle is still debatable.

25. Hur, Y. M., Bouchard Jr., T. J., and Eckert, E. (1998). Genetic and environmental influences on self-reported diet: A reared-apart twin study. *Physiology & Behavior, 64,* 629–636.

Of course, the approximately 30 percent heritable contribution to food preference reflects all heritability, not just variation in those genes involved in taste sensation.

26. Smith, A. D., et al. (2016). Genetic and environmental influences on food preferences in adolescence. *American Journal of Clinical Nutrition, 104,* 446–453.

27. For an enlightening tour of the multisensory nature of dining and fun anecdotes about fancy modern restaurants, I recommend: Spence, C. (2018). *Gastrophysics: The new science of eating.* New York, NY: Viking.

28. This also means that if you are tasting wines by the swish-and-spit method, you really should try to hold the wine in your mouth while exhaling to get a fuller effect. In truth, however, swish and spit can never truly replace swallowing as a tasting method for wine or beer because there are bitter receptors located in the back of the throat that can only be activated by swallowing (or, God forbid, gargling).

29. We can only smell chemicals with a molecular weight lower than about 350 (by comparison, a single carbon atom has molecular weight of about 12). That said, just because a molecule is small and volatile and can get up your nose doesn't mean that we will be able to smell it. We can't smell CO_2 for example, but mosquitos and some other insects can because they have a specialized receptor for it.

30. There a whole layer of olfactory processing in the olfactory bulb that I'm leaving out here. If you're interested in exploring the olfactory bulb circuitry, a good place to start is in this fine textbook:

Luo, L. (2016). *Principles of neurobiology* (pp. 217–218). New York, NY: Garland Science.

31. Sosulski, D. L., Bloom, M. L., Cutforth, T., Axel, R., & Datta, S. R. (2011). Distinct representations of olfactory information in different cortical centres. *Nature, 472,* 213–216.

32. Laska, M. (2017). Human and animal olfactory abilities compared. In A. Buettner (Ed.), *Springer handbook of odor* (pp. 675–689). Basel, Switzerland: Springer International.

33. Porter, J., et al. (2006). Mechanisms of scent-tracking in humans. *Nature Neuroscience, 10,* 27–29.

To see a great photo of a subject in this experiment, see: Miller, G. (2006, December 18). Human scent tracking nothing to sniff at. *Science.* Retrieved from www.sciencemag.org/news/2006/12/human-scent-tracking-nothing-sniff.

34. You'll recall that dolphins and whales lack sweet, sour, bitter, and umami taste. This suggests that maybe whales would lack a sense of smell as well, but this is not true, at least for a handful of whale species studied to date. Why should dolphins and whales lack most taste sensations, but whales retain smell while dolphins have lost it? We don't know.

35. Scholz, A. T., Horrall, R. M., Cooper, J. C., & Hasler, A. D. (1976). Imprinting to chemical cues: The basis for home stream selection in salmon. *Science, 192,* 1247–1249.

36. Gilad, Y., Wiebe, V., Przeworski, M., Lancet, D., & Paabo, S. (2004). Loss of olfactory receptor genes coincides with the acquisition of full trichromatic vision in primates. *PLoS Biology, 2,* e5.

The authors of this paper appropriately caution the reader that the temporal coincidence of functional olfactory receptor gene loss with the emergence of trichromatic vision does not prove a causal relationship between these two changes.

37. Young, B. D. (2017). Smell's puzzling discrepancy: Gifted discrimination yet pitiful identification. *Mind & Language, 2019,* 1–25.

These percentages are for object recognition by orthonasal smell. The situation for retronasal smell may be somewhat different.

38. An obvious class of explanation for the frequent failure to name familiar odors is that neural connections between brain regions responsible for odor detection and brain regions responsible for storing the names of objects are weak or circuitous. As one example, it has been pointed out that, unlike other sensory information, olfactory information does not get routed to the thalamus, a brain structure important for processing and distributing sensory signals. While this is true, it is not clear that the lack of direct odor input to the thalamus is related to the failure of naming familiar odors.

39. This doesn't mean that one can't have source-based color descriptors. For example, before the introduction of the fruit called "orange" to England, there was no specific name for the corresponding color. It was called yellow-red.

40. Olofsson, J. K., & Gottfried, J. A. (2015). The muted sense: Neurocognitive limitations of olfactory language. *Trends in Cognitive Science, 19,* 314–321.

41. Dupire, M. (1987). Des goûts et des odeur: Classifications et universaux. *L'Homme, 27,* 5–25.

42. Majid, A. (2015). Cultural factors shape olfactory language. *Trends in Cognitive Science, 19,* 629.

43. Wnuk, E., & Majid, A. (2014). Revisiting the limits of language: The odor lexicon of Maniq. *Cognition, 131,* 125–138.

Speakers of Jahai and Maniq also sometimes use source-based descriptors for odors, but it's rare.

44. Majid, A., & Burenhult, N. (2014). Odors are expressible in language, as long as you speak the right language. *Cognition, 130,* 266–270.

45. Bosker writes:

I added to a running list of the esoteric things people came up with after sticking their noses in each glass. It sounded like they were reading recipes from a Wiccan book of love spells: "wild strawberry water," "dry and rehydrated black fruits," "apple blossom," "saffron lobster stock," "burnt hair," "decomposing log," "jalapeño skin," "old aspirin," "baby's breath," "sweat," "chocolate-covered mint," "spent ground coffee," "confected violet," "strawberry fruit leather," "pleather," "freshly molded dildo," "horse tack," "dusty road," "lemon rind," "nail-polish remover," "stale beer," "fresh-tilled earth," "red forest floor," "pear drops," "cowhide," "desiccated strawberry," and "Robitussin."

Bosker, B. (2017). *Cork dork: A wine-fueled adventure among the obsessive sommeliers, big bottle hunters, and rogue scientists who taught me to live for taste* (pp. 199–200). New York, NY: Penguin.

46. Livermore, A., & Liang, D. G. (1996). Influence of training and experience on the perception of multicomponent odor mixtures. *Journal of Experimental Psychology: Human Perception and Performance, 22,* 267–277.

47. Morrot, G., Brochet, F., & Dubourdieu, D. (2001). The color of odors. *Brain and Language, 79,* 309–320.

48. Slosson, E. E. (1899). A lecture experiment in hallucinations. *Psychological Review, 6,* 407–408.

49. O'Mahony, M. (1978). Smell illusions and suggestion: Reports of smells contingent on tones played on television and radio. *Chemical Senses and Flavour, 3,* 183–189.

50. Although only sixteen people wrote in to report "no smell," it is likely that fewer people who failed to experience a smell were motivated to write.

51. Campenni, C. E., Crawley, E. J., & Meier, M. E. (2004). Role of suggestion in odor-induced mood change. *Psychological Reports, 94,* 1127–1136.

52. Herz, R. S., & von Clef, J. (2001). The influence of verbal labeling on the perception of odors: Evidence for olfactory illusions? *Perception, 30,* 381–391.

For more on olfactory illusions and olfactory learning, see Rachel Herz's fine book:

Herz, R. (2007). *The scent of desire: Discovering our enigmatic sense of smell.* New York, NY: HarperCollins.

53. Many of the intrinsically attractive or aversive odors are detected by a small subset of special odorant receptors called trace-amine associated receptors (TAARs). Humans have five functional receptors of this family, while mice have fourteen. The odor trimethylamine, which is aversive to rats and humans but attractive to mice, is detected by the receptor called TAAR5. For more on innate odor responses in various animals, see:

Li, Q., & Liberles, S. D. (2015). Aversion and attraction through olfaction. *Current Biology, 25,* R120–R129.

54. Wasabi, horseradish, and yellow mustard all contain a chemical called allyl isothiocyanate, which activates the receptor TRPA1. Another compound, diallyl disulfide, found in raw onions and garlic, also activates TRPA1 and so evokes a warming sensation. It's interesting that different families of plants have independently evolved the ability to produce compounds that activate TRPA1. If you'd like to read more about chemical and temperature sensing, see:

Linden, D. J. (2015). *Touch: The science of hand, heart and mind* (pp. 122–142). New York, NY: Viking.

55. Keller, A., Hempstead, M., Gomez, I. A., Gilbert, A. N., & Vosshall, L. B. (2012). An olfactory demography of a diverse metropolitan population. *BMC Neuroscience, 13,* 122.

56. Sorokowski, P., et al. (2019). Sex differences in human olfaction: A meta-analysis. *Frontiers in Psychology, 10,* 242.

While, on average, women have lower detection thresholds for odors then men, there are some individual odors for which this trend is reversed. For example, men tend to be more sensitive to the chemical bourgeonal, which has a lily of the valley scent. Why? We have no idea.

57. Sorokowska, A. (2016). Olfactory performance in a large sample of early-blind and late-blind individuals. *Chemical Senses, 41,* 703–709.

58. Wysocki, C. J., & Gilbert, A. N. (1989). National Geographic smell survey. Effect of age are heterogenous. *Annals of the New York Academy of Sciences, 561,* 12–28.

59. Keller, A., Zhuang, H., Chi, Q., Vosshall, L. B., & Matsunami, H. (2007). Genetic variation in a human odorant receptor alters odor perception. *Nature, 449,* 468–472.

60. Trimmer, C., et al. (2019). Genetic variation across the human olfactory receptor repertoire alters odor perception. *Proceedings of the National Academy of Sciences of the USA, 116,* 9475–9480.

61. Markt, S. C., et al. (2016). Sniffing out significant "pee values": Genome wide association study of asparagus anosmia. *BMJ, 355,* i6071.

None of these mutations were smack in the middle of the coding region of a particular olfactory receptor gene, but rather were in close proximity to genes called *OR2M7*, *OR2L3*, and *OR14C36*. There are many examples of mutations near the coding regions of genes that affect their expression, even if the structure of the encoded protein is unaltered.

62. Wysocki, C., Dorries, K. M., & Beauchamp, G. K. (1989). Ability to perceive androstenone can be acquired by ostensibly anosmic people. *Proceedings of the National Academy of Sciences of the USA, 86,* 7976–7978.

63. And the story gets even more complicated. Another group has shown that repeated exposure to benzaldehyde, which smells like almonds, or citralva, which smells lemony, can lead to lowered olfactory threshold for this odorant, but only in women and only when the women are of reproductive age. The size of this effect is large, about a hundred-thousand-fold reduction in detection threshold, rivaling the olfactory capabilities of dogs.

Dalton, P., Doolittle, N., & Breslin, P. A. S. (2002). Gender-specific induction of enhanced sensitivity to odors. *Nature Neuroscience, 5,* 199–200.

Diamond, J., Dalton, P., Doolittle, N., & Breslin, P. A. S. (2005). Gender-specific olfactory sensitization: Hormonal and cognitive influences. *Chemical Senses, 30,* i225–i225.

64. Wang, L., Chen, L., & Jacob, T. (2003). Evidence for peripheral plasticity in human odour response. *Journal of Physiology, 554,* 236–244.

65. Ibarra-Sora, X., et al. (2017). Variation in olfactory neuron repertoires is genetically controlled and environmentally modulated. *eLife, 6,* e21476.

66. Mainland, J. D., et al. (2002). One nostril knows what the other learns. *Nature, 419,* 802.

67. Alternatively (or in addition), a more complicated explanation is that the brain serves as a conduit that can pass signals from one nostril to the other to render the unexposed side of the nose more sensitive.

68. Gottfried, J. A., & Wu, K. N. (2009). Perceptual and neural pliability of odor objects. *Annals of the New York Academy of Science, 1170,* 324–332.

69. Royet, J. P., Plailly, J., Saive, A. L., Veyrac, A., & Delon-Martin, C. (2013). The impact of expertise in olfaction. *Frontiers in Psychology, 4,* 928.

70. Spahn, J. M., et al. (2019). Influence of maternal diet on flavor transfer to amniotic fluid and breast milk and children's responses: A systematic review. *American Journal of Clinical Nutrition, 109,* 1003S–1026S.

A similar but somewhat less robust effect on infants' food preferences is seen for the flavors of foods consumed by breastfeeding mothers.

71. Nguyen, D. H., Valentin, D., Ly, M. H., Chrea, C., & Sauvageot, F. (2002). *When does smell enhance taste? Effect of culture and odorant/tastant relationship.* Paper presented at the European Chemoreception Research Organisation conference, Erlangen, Germany.

72. Stevenson, R. J., Prescott, J., & Boakes, R. A. (1999). Confusing tastes and smells: How odors can influence the perception of sweet and sour tastes. *Chemical Senses, 24,* 627–635.

Stevenson, R. J., & Boakes, R. A. (1998). Changes in odor sweetness resulting from implicit learning of a simultaneous odor-sweetness association: An example of learned synesthesia. *Learning and Motivation, 29,* 113–132.

73. Cain, W. S., & Johnson Jr., F. (1978). Lability of odor pleasantness: Influence of mere exposure. *Perception, 7,* 459–465.

74. Moncreiff, R. W. (1966). *Odour preferences*. New York, NY: Wiley.

Wintergreen was ranked eighty-second out of 132 odors tested. The top-ranked odor was rose essence and the last-ranked odor was thiomalic acid, which I'm told smells like burned rubber.

75. Classen, C., Howes, D., & Synnott, A. (1994). *Aroma: The cultural construction of smell.* London: Routledge.

Chapter Seven. Sweet Dreams Are Made of This

1. For those of you too young to have encountered one, the Magic Fingers was a gizmo, common in motels in the 1960s and '70s, where you dropped in a quarter to activate a motor that would then vibrate the bed frame for fifteen minutes. The text on the side of the device read, "It quickly carries you into the land of tingling relaxation and ease."

2. Fischer, D., Lombardi, D. A., Marucci-Wellman, H., & Roenneberg, T. (2017). Chronotypes in the US—influence of age and sex. *PLoS One, 12,* e0178782.

These survey findings were mostly similar to previous reports in other economically developed countries like Germany and New Zealand.

3. Vetter, C., et al. (2015). Mismatch of sleep and work timing and risk of type 2 diabetes. *Diabetes Care, 38,* 1707–1713.

4. Walch, O. J., Cochran, A., & Forger, D. B. (2016). A global quantification of "normal" sleep schedules using smartphone data. *Science Advances, 2,* e1501705.

This study used a smartphone app for reporting. Importantly, it was not limited to weekend nights, and so cannot be directly compared with the findings in Fischer et al. The authors found that, mostly, those countries with an early average bedtime also had an early average waking time, and those with an average later bedtime also tended to wake later, suggesting that, roughly speaking, sleep duration is protected across these cultures.

5. Ekirch, A. R. (2001). Sleep we have lost: Pre-industrial slumber in the British Isles. *American Historical Review, 106,* 343–386.

6. Ekirch, A. R. (2016). Segmented sleep in pre-industrial societies. *Sleep, 39,* 715–716.

7. Yetish, G., et al. (2015). Natural sleep and its seasonal variations in three pre-industrial societies. *Current Biology, 25,* 2862–2868.

8. De la Iglesia, H. O., et al. (2015). Access to electric light is associated with shorter sleep duration in a traditionally hunter-gatherer community. *Journal of Biological Rhythms, 30,* 342–350.

9. Pilz, L. K., Levandovski, R., Oliveira, M. A. B., Hidalgo, M. P., & Roenneberg, T. (2018). Sleep and light exposure across different levels of urbanization in Brazilian communities. *Scientific Reports, 8,* 11389.

10. Yetish, G., et al. (2015). Natural sleep and its seasonal variations in three pre-industrial societies. *Current Biology, 25,* 2862–2868.

De la Iglesia, H. O., et al. (2015). Access to electric light is associated with shorter sleep duration in a traditionally hunter-gatherer community. *Journal of Biological Rhythms, 30,* 342–350.

11. Ekirch, A. R. (2016). Segmented sleep in pre-industrial societies. *Sleep, 39,* 715–716.

12. This factoid comes from a wonderful book about biological rhythms: Foster, R., & Kreitzman, L. (2004). *Rhythms of life: The biological clocks that control the daily lives of every living thing.* London: Profile Books.

13. For those of you who are curious about the details of the circadian clock, I recommend these two review articles:

Bedont, J. L., & Blackshaw, S. (2015). Constructing the suprachiasmatic nucleus: A watchmaker's perspective. *Frontiers in Systems Neuroscience, 9,* 74.

Takahashi, J. S. (2017). Transcriptional architecture of the mammalian circadian clock. *Nature Reviews Genetics, 18,* 164–179.

14. Importantly, not only are intrinsically photosensitive ganglion cells stimulated by strong sunlight, but they can also be activated by relatively weak artificial lighting, particularly if it contains blue photons. As a result, when you stay up late under artificial light, including that from your smartphone or tablet screen, you are trying to force your internal circadian clock into a longer period, disrupting natural rhythms.

15. Jones, C. R., et al. (1999). Familial advanced sleep-phase syndrome: A short-period circadian rhythm variant in humans. *Nature Medicine, 5,* 1062–1065.

16. They are a married couple.

17. Toh, K. L., et al. (2001). An *hPer2* phosphorylation site mutation in familial advanced sleep phase syndrome. *Science, 291,* 1040–1043.

18. Shi, G., Wu, D., Ptáček, L. J., & Fu, Y. F. (2017). Human genetics and sleep behavior. *Current Opinion in Neurobiology, 44,* 43–49.

19. Shi, G., et al. (2019). A rare mutation of β1-adrenergic receptor affects sleep/wake behaviors. *Neuron, 103,* 1–12.

20. Funato, H., et al. (2016). Forward-genetics analysis of sleep in randomly mutagenized mice. *Nature, 539,* 378–383.

Hayasaka, N., et al. (2017). Salt-inducible kinase 3 regulates the mammalian circadian clock by destabilizing Per2 protein. *eLife, 6,* e24779.

21. Gehrman, P. R., et al. (2019). Twin-based heritability of actimetry traits. *Genes, Brain and Behavior, 18,* e12569.

22. Kalmbach, D. A., et al. (2017). Genetic basis of chronotype in humans: Insights from three landmark GWAS. *Sleep, 40,* 1–10.

Jones, S. E., et al. (2019). Genome-wide association analysis of chronotype in 697,828 individuals provides insight into circadian rhythms. *Nature Communications, 10,* 343.

23. Michel Jouvet of the University of Lyon showed in cats that severing the inhibitory fibers that block motor outflow resulted in odd behavior: During REM sleep they engaged in complex movements, all while keeping their eyes closed. They ran, pounced, and even appeared to eat their imagined prey.

24. Mahoney, C. E., Cogswell, A., Koralnik, I. J., & and Scammell, T. E. (2019). The neurobiological basis of narcolepsy. *Nature Reviews Neuroscience, 20,* 83–93.

25. This case report comes from: Dauvilliers, Y., & Barateau, L. (2017). Narcolepsy and other central hypersomnias. *Continuum, 23,* 989–1004.

26. Mignot, E. (1998). Genetic and familial aspects of narcolepsy. *Neurology, 50,* S16–S22.

27. Loss of the orexin neurons is key to narcolepsy, but it may not be the whole story. It's possible that the loss of some or all orexin neurons triggers compensatory responses in other brain systems, and those compensatory responses may be helpful in some cases but harmful in others. Presently, there's no way to restore the lost orexin neurons in the brain, so treatment of narcolepsy is symptom based: usually stimulants like modafinil for daytime sleepiness and SSRIs for cataplexy.

28. Stickgold, R., et al. (2000). Replaying the game: Hypnagogic images in normals and amnesiacs. *Science, 290,* 350–353.

29. Solms, M. (2000). Dreaming and REM sleep are different. *Behavioral and Brain Sciences, 23,* 793–1121.

30. Nir, Y., & Tononi, G. (2009). Dreaming and the brain: From phenomenology to neurophysiology. *Trends in Cognitive Science, 14,* 88–100.

31. Dement, W., & Wolpert, E. A. (1958). The relation of eye movements, body motility and external stimuli to dream content. *Journal of Experimental Psychology, 55,* 543–553.

32. Rechtschaffen, A., & Foulkes, D. (1965). Effect of visual stimuli on dream content. *Perceptual and Motor Skills, 20,* 1149–1160.

33. Butler, S., & Watson, R. (1985). Individual differences in memory for dreams: The role of cognitive skills. *Perceptual and Motor Skills, 53,* 841–964.

34. Nir, Y., & Tononi, G. (2009). Dreaming and the brain: From phenomenology to neurophysiology. *Trends in Cognitive Science, 14,* 88–100.

35. De Gannaro, L., et al. (2016). Dopaminergic system and dream recall: An MRI study in Parkinson's disease patients. *Human Brain Mapping, 37,* 1136–1147.

36. Snyder, F. (1970). The phenomenology of dreaming. In L. Madow and L. H. Snow (Eds.), *The psychodynamic implications of the physiological studies on dreams* (pp. 124–151). Springfield, IL: Charles C. Thomas.

Hall, C., & Van de Castle, R. (1966). *The content analysis of dreams.* New York, NY: Appleton-Century-Crofts.

Domhoff, G. W. (2003). *The scientific study of dreams: Neural networks, cognitive development, and content analysis*. Washington, DC: American Psychological Association.

Foulkes, D. (1985). *Dreaming: A cognitive-psychological analysis*. Hillsdale, NJ: Lawrence Erlbaum Associates.

Chapter Eight. A Day at the Races

1. Sather, C. (1997). *The Bajau Laut: Adaptation, history, and fate in a maritime fishing society of south-eastern Sabah.* Kuala Lumpur, Malaysia: Oxford University Press.

2. Ivanoff, J. (1997). *Moken: Sea-gypsies of the Andaman Sea—Post-war chronicles.* Chonburi, Thailand: White Lotus Press.

3. Schagatay, E., Losin-Sundström, A., & Abrahamsson, E. (2011). Underwater working times in two groups of traditional apnea divers in Asia: The Ama and the Bajau. *Diving and Hyperbaric Medicine, 41,* 27–30.

4. Schagatay, E. (2014). Human breath-hold diving and the underlying physiology. *Human Evolution, 29,* 125–140.

5. Ilardo, M., et al. (2018). Physiological and genetic adaptations to diving in sea nomads. *Cell, 173,* 569–580.

6. Gislén, A., et al. (2003). Superior underwater vision in a human population of sea gypsies. *Current Biology, 13,* 833–836.

7. Gislén, A., Warrant, E. J., Dacke, M., & Kröger, R. H. H. (2006). Visual training improves underwater vision in children. *Vision Research, 46,* 3443–3450.

8. Fan, S., Hansen, M. E. B., Lo, Y., & Tishkoff, S. A. (2016). Going global by adapting local: A review of recent human adaptation. *Science, 354,* 54–59.

Ilardo, M., & Nielsen, R. (2018). Human adaptation to extreme environmental conditions. *Current Opinion in Genetics & Development, 53,* 77–82.

9. At present, it is rare to recover ancient DNA from bones in most parts of Africa and other warm places, because the DNA is degraded in that climate. As a result, our understanding of how lactase persistence spread in Africa is quite limited. In addition to lactase persistence, the ability of adults to metabolize milk can be influenced by the population of microbes in the gut.

10. In some cases, we know that different gene variants selected in hypoxic environments can be related in terms of biochemical or genetic signaling. For example, both *EGLN1* and *EPAS1* act in the same signaling pathway to activate the hypoxia-inducible transcription factor HIF.

11. A twist to this story comes from the tropical rainforests of western Central Africa, where there live people of unusually short stature, called pygmies. Short stature seems to be a common adaptation in rainforest hunter-gatherer populations, as it is also seen in the Amazon basin and in Southeast Asia. It's not really clear why short stature should be adaptive in tropical rainforest hunter-gatherer people. One interesting possibility is that short stature is not an adaptation at all but is merely a byproduct of speeding up postnatal development. The hypothesis is that in high-mortality environments, it is adaptive to develop quickly and have your children early before you are likely to die.

Migliano, A. B., Vinicius, L., & Lahr, M. M. (2007). Life history trade-offs explain the evolution of human pygmies. *Proceedings of the National Academy of Sciences of the USA, 104,* 20216–20219.

Short stature is a highly heritable trait among all of these pygmy groups and is not merely a result of widespread malnutrition. When DNA from pygmy groups living in Cameroon was analyzed, it was found that their short stature could be accounted for by variation in only a small number of genes, one of which influenced the production of growth hormone. In this way, height turns out to be an interesting trait. In most people around the world, variation in height comes from tiny effects of variation in hundreds of genes. But in African pygmies, who make up a miniscule fraction of the world's human population, this highly polygenic process has been overridden by the strong action of a small number of genes to create unusually short stature.

Jarvis, J. P., et al. (2012). Patterns of ancestry, signatures of natural selection, and genetic association with stature in Western African pygmies. *PLoS Genetics, 3,* e1002641.

12. Turchin M. C., et al. (2012). Evidence of widespread selection on standing variation in Europe at height-associated SNPs. *Nature Genetics, 44,* 1015–1019.

13. Field, Y., et al. (2016). Detection of human adaptation during the past 2,000 years. *Science, 354,* 760–764.

14. Sohail, M., et al. (2019). Polygenic adaptation on height is overestimated due to uncorrected stratification in genome-wide population studies. *eLife, 8,* e39702.

Berg, J. J., et al. (2019). Reduced signal for polygenic adaptation of height in UK Biobank. *eLife, 8,* e39725.

15. Chabris, C. F., Lee, J. J., Cesarini, D., Benjamin, D. J., & Laibson, D. I. (2015). The fourth law of behavior genetics. *Current Directions in Psychological Science, 24,* 304–312.

16. Saini, A. (2019). *Superior: The return of race science.* Boston, MA: Beacon Press.

17. For a clear and compelling refutation of such fake evolutionary arguments, see:

Rutherford, A. (2020). *How to argue with a racist: What genes do (and don't) say about human difference*. New York, NY: The Experiment.

18. See: About Race. (2018, January 23). United States Census Bureau. Retrieved from: www.census.gov/topics/population/race/about.html.

19. Norton, H. L., Quillen, E. E., Bigham, A. W., Pearson, L. N., & Dunsworth, H. (2019). Human races are not like dog breeds: Refuting a racist analogy. *Evolution: Education and Outreach, 12,* 17.

20. Marks, J. (2017). *Is science racist?* (pp. 53–54). Cambridge, UK: Polity Press.

21. For some historical perspective, the first animal gene was cloned from the clawed frog *Xenopus* in 1973, a year after Lewontin's report, and the entire human genome was not sequenced until 2003.

22. Lewontin, R. C. (1972). The apportionment of human diversity. *Evolutionary Biology, 6,* 381–398.

23. Rosenberg, N. A., et al. (2002). Genetic structure of human populations. *Science, 298,* 2381–2385.

Rosenberg, N. A., et al. (2005). Clines, clusters, and the effect of study design on the inference of human population structure. *PLoS Genetics, 1,* e70.

The 2002 paper was critiqued on the basis that the clusters observed were in large part an artifact of nonrandom geographic sampling of these populations. The 2005 paper largely addressed those concerns.

These authors also identified a sixth cluster specific to the Kalash people, a population in Pakistan, likely reflecting a high degree of inbreeding and genetic drift in that group.

24. Tischkoff, S. A., et al. (2009). The genetic structure and history of Africans and African Americans. *Science, 324,* 1035–1044.

25. Tischkoff, S. A., & Kidd, K. K. (2004). Implications of biogeography of human populations for "race" and medicine. *Nature Genetics, 36,* S21–S27.

26. In many cases, the ancient DNA evidence has reinforced the evidence from archeology, which has also argued against concepts of historical racial purity.

27. Reich, D. (2018). *Who we are and how we got here: Ancient DNA and the new science of the human past.* New York, NY: Pantheon Books.

28. When you spit in a tube and mail off your DNA sample, the direct-to-consumer ancestry company is not analyzing all three billion or so nucleotides of your DNA. Rather, they are analyzing an array of about six hundred thousand to one million nucleotides that are known to be variable across humans. Then this pattern is compared with DNA samples from people around the world that have been chosen to have so-called unmixed recent ancestry. For example, for their Greek samples, they would pick people for whom all four grandparents lived in Greece. There are several sources of error in this process. First, the step by which the DNA from your spit is amplified prior to analysis can introduce errors. That's why even identical twins can get back ancestry reports from the same company that are not exactly the same. Second, the database depends on accurate reporting about the grandparents of the reference subjects, and this is not always possible (maybe that Irish grandma was fooling around with her Polish neighbor and not telling her Irish husband). Third, these estimates are better for populations where there are a lot of reference samples and poorer for populations that are sparsely represented in the database.

29. Gottfredson, L. S. (1997). Mainstream science on intelligence: An editorial with 52 signatories, history and bibliography. *Intelligence, 24,* 13–23.

30. Ree, M. J., & Earles, J. A. (1991). The stability of *g* across different methods of estimation. *Intelligence, 15,* 271–278.

Nisbett, R. E., et al. (2012). Intelligence: New findings and theoretical developments. *American Psychologist, 67,* 130–159.

31. Interestingly, intellectual disability, defined as an IQ score less than or equal to seventy, is about fourfold more prevalent in men than in women. One likely explanation for this is that having a single X chromosome leaves males more vulnerable to X-linked mutations that affect brain development or plasticity. Indeed, over one hundred such mutations on the X chromosome have been found to date.

Carvill, G. L., & Mefford, H. C. (2015). Next-generation sequencing in intellectual disability. *Journal of Pediatric Genetics, 4,* 128–135.

32. Most twin studies have an unintentional economic class bias built into them, as people from lower socioeconomic classes are less likely to come to the lab and participate. This probably accounts for the somewhat higher estimates of heritability in IQ from twin studies as compared with other methods.

33. Zogbhi, H. Y., & Bear, M. F. (2012). Synaptic dysfunction in neurodevelopmental disorders associated with autism and intellectual disabilities. *Cold Spring Harbor Perspectives in Biology, 4.* doi:10.1101/cshperspect.a009886.

34. Eysenck, H. J. (1971). *The IQ argument: Race, intelligence and education.* New York, NY: Library Press.

35. This example is from: Mitchell, K. (2018) *Innate: How the wiring of our brains shapes who we are.* Princeton, NJ: Princeton University Press.

36. Lynn, R., & Vanhanen, T. (2012). National IQs: A review of their educational, cognitive, economic, political, demographic, sociological epidemiological, geographic and climactic correlates. *Intelligence, 40,* 226–234.

Carl, N. (2016). IQ and socioeconomic development across regions of the UK. *Journal of Biosocial Science, 48,* 406–417.

37. Scarr-Salapatek, S. (1971). Race, social class and IQ. *Science, 174,* 1285–1295.

Rowe, D. C., Jacobson, K. C., & Van den Oord, E. J. (1999). Genetic and environmental influences on vocabulary IQ. *Child Development, 70,* 1151–1162.

Turkheimer, E., Haley, A., Waldron, M., D'Onofrio, B., & Gottesman, I. I. (2003). Socioeconomic status modifies heritability of IQ in young children. *Psychological Science, 14,* 623–628.

38. Dickens, W. T., & Flynn, J. R. (2006). Black Americans reduce the racial IQ gap: Evidence from standardization samples. *Psychological Science, 17,* 913–920.

39. Hart, B., & Risley, T. (1995). *Meaningful differences in the everyday experience of young American children.* Baltimore, MD: Paul H. Brookes Publishing.

40. Locurto, C. (1990). The malleability of IQ as judged from adoption studies. *Intelligence, 14,* 275–292.

Duyme, M., Dumaret, A., & Tomkiewicz, S. (1999). How can we boost the IQs of "dull" children? A late adoption study. *Proceedings of the National Academy of Sciences of the USA, 96,* 8790–8794.

Van IJzendoorn, M. H., Juffer, F., & Poelhuis, C. W. K. (2005). Adoption and cognitive development: A meta-analytic comparison of adopted and non-adopted children's IQ and school performance. *Psychological Bulletin, 131,* 301–316.

41. Zogbhi, H. Y., & Bear, M. F. (2012). Synaptic dysfunction in neurodevelopmental disorders associated with autism and intellectual disabilities. *Cold Spring Harbor Perspectives in Biology, 4.* doi:10.1101/cshperspect.a009886.

42. Davies, G., et al. (2015). Genetic contributions to variation in general cognitive function; a meta-analysis of genome-wide association studies in the CHARGE consortium (N=53,949). *Molecular Psychiatry, 20,* 183–192.

Savage, J. E., et al. (2018). Genome-wide association meta-analysis in 269,867 individuals identifies new genetic and functional links to intelligence. *Nature Genetics, 50,* 912–919.

43. At present, it's not possible to look at the thousand-plus intelligence-linked gene variants and derive a common neurophysiological function for those genes that underlies improved intelligence. People have speculated that intelligence is promoted by neural ensembles that process information more efficiently (using less energy per unit of information) or by having neurons with bigger dendrites (the tree-shaped information-receiving ends), but these reasonable hypotheses remain unproven.

44. This holds for both the thousand-plus genes that form the polygenic basis for the heritable portion of IQ test score and the smaller number of genes in which a mutation can result in intellectual disability.

Epilogue

1. Willoughby, E. A., et al. (2019). Free will, determinism and intuitive judgments about the heritability of behavior. *Behavior Genetics, 49,* 136–153.

The sample of this online survey of adults in the United States was likely biased toward more educated and higher socioeconomic class individuals. Interestingly, the subgroup of the survey population with the best average match between their trait heritability estimates and the consensus from the scientific literature was educated mothers with multiple children.

2. Ota, I., et al. (2010). Association between breast cancer risk and the wild-type allele of human ABC transporter *ABCC11*. *Anticancer Research, 30,* 5189–5194.

Index

NATALIE LINDEN

David J. Linden is a professor of neuroscience at the Johns Hopkins University School of Medicine, where he studies the molecular substrates of memory storage and recovery after brain injury. He is the author of three books: *Touch*, *The Accidental Mind*, and *The Compass of Pleasure*. He lives in Baltimorc, Maryland.